# Pasaquina

a novel of
El Salvador

by

Erin O'Shaughnessy

Saybrook

Publishers

The village of Pasaquina does exist in El Salvador. The characters and events depicted in this novel, however, are entirely fictitious.

The epigraph from the book of Ecclesiastes is quoted from *The Holy Bible,* translated by Ronald Knox. New York: Sheed & Ward, Inc., 1956. Copyright © 1950 by Sheed and Ward. Reprinted by permission.

Library of Congress Cataloging-in-Publication Data

O'Shaughnessy, Erin, 1942—
    Pasaquina.

    I. Title.
PS3565.S54P37   1986        813'.54        86-6703
ISBN 0-933071-05-1

Saybrook Publishers
4223 Cole Avenue, Suite Four, Dallas, TX 75205

Printed in the United States of America

Distributed by W.W. Norton & Company
500 Fifth Avenue, New York, NY 10110

After a while it becomes evident that it is time to stop the killing. Some of you know that. This story is for you.

...a shadow's shadow; a world of shadows!
How is man better for all this toiling of his, here
under the sun? Age succeeds age, and the world
goes on unaltered. Sun may rise and sun may set,
but ever it goes back and is reborn. Round to the
south it moves, round to the north it turns; the
wind, too, though it makes the round of the world,
goes back to the beginning of its round at last. All
the rivers flow into the sea, yet never the sea
grows full; back to their springs they find their
way, and must be flowing still. Weariness, all
weariness; who shall tell the tale?

Ecclesiastes 1.2-8,
translated by Ronald Knox

No Latino villager ever hurries, for, after all, where is there to go? They are either going to heaven or to hell, and that has already been decided. Only the Anglos hurry, and Father Herrera says that is because they are trying to live like hell on earth while at the same time planning how to cheat God into going to heaven at the last minute.

So when the boys came down from the mountain for their Saturday confessions, they came lazily, as if they were wandering in from the fields for a market day. But appearances are, of course, deceptive. In reality, they had crawled out of their brush lean-tos and cedar-scented caves to descend on a pilgrimage, the object of which was to keep their souls clean in preparation for a quick exit. It was a regular Saturday ritual, weather and war permitting, and the villagers all knew this. What with some of the villagers having cousins and uncles and fathers in the army because it paid better than farming, the army, and probably the Guardia Nacional, knew about the confessions of los muchachos too, but they never made any attempt to interfere. No ambushes were ever set or traps sprung; such behavior would have been considered despicable, not to mention dishonorable. It would have been thought cowardly, like killing the bull with a sword thrust to the lungs.

Should one side or the other get a reputation for cowardice, all agreed, that would finish the war, and nobody was entirely sure that they wanted that, for the finish of the war would mean the finish of American doctors and their medicines and American rice and American beans and surplus shoes and the good business the country was having from the black market in stolen American merchandise. So finishing the war was a doubtful subject, fit only for argumentation in the cafés at siesta time, not a serious issue.

Then too, an attack on los muchachos would have been vicious, not at all the sort of thing done here in the countryside. Rumor had it that far away in San Salvador, atrocities and crimes were served up every morning along with the pan francés and café con leche. But this was hard to believe, and in any case, could never be true in Pasaquina. So the boys wound down the mountainside unmolested, and made their way to the village priest.

Father Herrera lived in a five-room, white-washed adobe house built around a flagstone courtyard behind the church, and he washed faithfully in a galvanized metal tub every Saturday morning before confessions. He was a short, round, brown man with a fondness for pig's lard and sour lime with his vino. When hearing confessions, he wore a once-black cassock which had faded to a purplish gray color and was so limp with wear that it clung to his thighs and buttocks like a woolen jersey. His sandals were too long for him, as he had abnormally short feet, and when he walked he lifted his toes high in the front so as not to trip himself on the extra leather of his soles. He considered the effect stately, but in fact he ended up resembling nothing so much as a small, squat penguin, with all the seriousness and none of the elegance. He was a very sincere man, however, devoted to his duty of saving souls, and entirely convinced that he was doing so.

Father Herrera considered the rituals of the Mass sacrosanct; indeed, he insisted on cleaning beneath his fingernails before handling the Host. And once, when the bishop was coming to visit, he had even had his criada steal the gold thread from an officer's epaulets to re-embroider the designs on the edges of his one good white stole. He considered it borrowing and fully meant to unravel the girl's work and return the threads after the bishop left town, but somehow, with one thing and another—deaths, burials, births, sicknesses, family quarrels and so on—it never got done; so when the officer came in to confess that he had taken a virgin (already betrothed to someone else) against her will, the padre went easy on the penances. Since the officer, like everyone else, had noticed the fresh shine on the fringes of the priest's stole when the bishop came to call, and therefore understood why he was so little chastised for almost, but not quite, raping the maiden, no questions were asked, and no answers given. The matter was considered settled.

Pasaquina suited Father Herrera, who found the lack of formality in village life easy on the soul and compatible with his unambitious nature. He had been there some little time, having been sent to replace a priest whose death was expected momentarily for nearly three years. In that time the village became accustomed to Father Herrera's shambling presence, and when the transition finally did take place, it was hardly noticeable. He at once replaced his predecessor's spinster maid with a young, plump criada whom he frequently pinched below the waistline, indiscriminately in the front or back, for he liked her rounded stomach as much as her rounded rump.

He himself taught the girl to cook so that he could enjoy all the pleasures of his rank. She had a separate room set apart from his quarters, and when she proved faithful and docile even after she had borne her first child by him, he sent all the way to San Salvador for a fine mirror for her, painted with gold scrollwork around the edges. Her mirror was the envy of the village, and the spinsters and widows and giggling girls paraded in by the back entrance, one by one, to see it. After they had done so, and talked it up throughout the village, the girl's position in the priest's household was established as permanent.

This arrangement satisfied everyone, and not a ripple or a stir was heard about it until several years later when the girl, whose name was Rita, had a brief encounter with a young guerrilla by the name of Manuel. There was a great amount of gossip and discussion then about the propriety of such a love affair, and whether or not the priest should be ordered to be rid of her, as perhaps she would be unclean. But some of the wiser men of the village concluded at last that it was only natural and in keeping with God's laws that a healthy, rosy-cheeked girl of nineteen would lust after the body of a strong, straight-toothed young man, and that the priest, being the authority on matters of God's laws, must surely know this better than they, and that was the explanation of his tolerance.

In fact, Father Herrera had failed entirely to notice the late night trysts of his criada and thought when her belly swelled with child that he had again made himself proud.

He was oblivious to the whole matter. But his cheerful ignorance—along with Pasaquina's tranquility—was threatened by the hot-blooded young Manuel, who, upon discovering that Rita bore his child, insisted that she come away with him and live in the mountains as his wife.

Rita wept and sobbed and argued and swore that she could not desert the priest, for then their child would surely be doomed to the eternal fires and damnation. Manuel accused her of softness and lack of honor, saying that she only wished to stay with the fat, greasy old man because he fed her well and provided her with a cotton mattress. She insisted it was not so and her only concern was for the soul of their unborn child, but he would not relent and stalked off into the hills an angry young man.

In his absence, Rita pined for him and developed a long, brooding face, which puzzled the priest and made him wonder if she had a cancer in her womb instead of a child, but in truth her unhappiness was only because she was considering relenting to her sweetheart's demands and was dreading it heartily, for winter was coming. The tension continued for several weeks, during which Manuel came twice to Saturday morning confession but refused to visit Rita's room afterward, and her soft brown face turned gray with the fatigue of rejection.

But the affair was at last ended by what the entire village considered an act of God. On the third Saturday of their power struggle, Manuel had relented in his anger, or grown overly hot and frustrated in his abstinence, and had paid his regular visit to Rita's room after his confession. Having finished up his business there, he drank from the priest's agua ardiente bottle, promised eternal fidelity to the only love of his life, and wandered under the moonlight back up the mountain trail toward his hideout. On the way, however, he stumbled on a hidden mine meant for any soldiers foolhardy enough to climb mountains drunkenly in the dark, and was blown into five or six parts.

The entire village agreed this was an excellent solution to the problem with the priest's criada and piously attended Manuel's funeral, dutifully wailing at the top of their lungs all the way to the cemetery. Appropriately, it rained, turning the dusty road into slippery clay, so that the men carrying Manuel's coffin slipped several times and even dropped the pine box once on an especially steep incline, which stimulated the stricken Rita to fall on it and afforded the old women of the village the opportunity to seize her by the neck and arms and shoulders and drag her off again so the journey could resume.

What with the rain and the wailing and the pregnant girl screaming and tearing her long black hair, and the fat, small priest unctuously reciting the words of the liturgy, it was a completely satisfactory experience, and every man who could afford it got drunk afterwards and raped his wife, and every woman who could afford it ate heartily afterwards and lamented aloud the fate of women everywhere long into the night.

With Manuel's passing, routine and tranquility returned once again to Pasaquina. In the morning the men climbed the slopes to the terraced fields of corn and sorghum while the women scrubbed clothes on the flat rocks at the stream bed, ground corn and spices on their piedra de molera or squatted on the low thresholds of their one-room houses to gossip and nurse infants. And life continued in the village square, where men gathered, dark and handsome in their hip-hugging pants, with machetes slung from their wide, dangerous-looking belts, drifting into little groups where they discussed the weather, the news, the feuds, the cock-fights, while sipping chibolas from chilled bottles or munching idly on sticks of cana.

Winter came, with only a passing chill at night, and though occasionally there was news of the war, it went primarily unnoticed, for Pasaquina was situated in a mountainous region so poor and far away from the cities that little action could be expected there. Although Pasaquina had once—only a decade ago—been prosperous, trading briskly with nearby Honduras, where cartloads of sombreros de palma, fine, crisp shirts, and sharp machetes were exchanged for cheese and sausage and goat's milk, that was before the Soccer War closed the border. The trading stopped and there was nothing left for Pasaquina but scratching the thin, blond earth. Since Pasaquina sat too high in the hills for the growing of coffee, perching lopsidedly astride a volcanic ridge where the soil had been washed away by the rainy seasons of many centuries, no colones were to be had from this grim earth, and so the oligarchy of coffee barons had no use for it. Consequently, their employees, the Guardia Nacional, had no use for it either.

With the approach of springtime, the orchid trees once more fell over the roads, the iguana came out to sun themselves on the rocks, and Rita gave birth to a small boy child. Once, a band of militia marched through the square, followed by a rattletrap jeep carrying officers of some sort, who announced themselves by the sternness of their countenances and the crossed straps of carbine cartridges on their chests. It caused a great stir at first but was soon forgotten, as no second band followed; so it was assumed that the war was only passing through and would not come to stay that year. And every Saturday morning the boys from the hills came down for their weekly confessions and a visit with their novias, if they had them. Occasionally, a restless one from the village would go back with them, sometimes to stay, sometimes to drift home after a few weeks, finding life on the mountain too hard or too monotonous to bear.

On special occasions, the guerrillas would march about the square in impressive parades to show their strength and to make their mothers weep and their sweethearts sigh. But the old men knew that this showing off was only possible because the Guardia was not interested in Pasaquina. Sometimes the guerrillas would brag about offensives which they said had occurred on the other side of the mountain, and once again the mothers would weep and the girls sigh and the old men refuse to believe them, spitting in the dust and snarling, "Then show me your scars, muchacho," calling them little boys to insult them.

But actually, the villagers were quite content to let los muchachos be "guerrillas" if that was what they wanted. There was, after all, little for young men to do, so in boredom and frustration they joined the army or went up the mountain to live like guerrillas. Either way was the same; it didn't matter. Some who tired of the discomforts of guerrilla life joined the army, and some in the army who were homesick deserted and returned home to join the guerrillas. It was all the same, and the war remained a distant matter, interesting, but not exactly important.

That spring a gringo came who said he was an American reporter, but no one believed him because he had a scrawny, knock-kneed girl with him—his photographer, he said—and they all agreed that anyone who could achieve the status of American reporter could surely do better as far as women were concerned, so he must be an imposter. For that reason they told him nothing but lies, and insisted as though with one voice that if only the Americans would send more food and money and medicine there would be no war, since they only fought because they were hungry. They refused to reveal to him the intricacies of the delicate, on-going debate over the relative merits of a dictatorship by the communist left or a dictatorship by the old-monied right, for they knew he could not possibly understand. The general consensus among the café society was that there was a great deal of good to be said about both propositions and a great deal of bad as well, and that life would be interminably boring without the possibility of struggle, and since only Latins could understand the subtleties of revolution, the visiting Americans were just children barging in on someone else's game. They suggested politely to the reporter that he should go home, or failing that, at least move to a more promising area. There were a number of suggestions about where that might be, but at last they all advised that he would be much happier in the capital, and that if he were lucky he might even take a bit of shrapnel in the shoulder there.

The reporter, of course, considering them hopelessly stupid, did exactly as they wished and went away. So once again life in Pasaquina was peaceful and predictable. During the winter, when the rains were heavy and the men were often confined to the doorsteps of their huts or—if they were rich—the awnings outside the cafés, it was discovered that the priest's criada, Rita, was again with child, and Father Herrera was searching the village for a widow woman to come in and help her with the work. The men shook their heads, saying too much money was going to Rita and not enough to the saints, and the women shook their heads and said Rita was a very stupid girl. They all sensed it would mean trouble and waited expectantly to hear about a visit from the bishop. But nothing happened, for the bishop, like everyone else of importance, had forgotten Pasaquina. So far as the rest of the world was concerned, it might not have existed at all.

So Pasaquina slumbered in peaceful oblivion all through that fall and the following winter. It was not until the next spring that two unprecedented occurrences, striking almost simultaneously, split the little village in two, as if a lightning bolt straight out of heaven had struck apart their iglesia and caused the mountain on which they lived to crack and fall in its entirety on their heads.

)

*Two* 🌙

The first occurrence was the one which, in the end, mattered the most, but which, in the beginning, seemed to matter the least. It was the arrival of Sister Magdalena. She came one bright summer morning in early August, riding on the back of a dusty gray burro which was being led by a small, barefoot boy. The boy's eyes, black as ink, winked out from under a mouse-chewed, wheat-colored straw sombrero. He had good, high Indian cheekbones and a round little mouth puckered up in whistling.

All that could be seen of Sister Magdalena were her delicate hands, holding lightly to the donkey's mane, and the tips of her toes beneath her robes. She rode sidesaddle on a faded red and green blanket, shrouded in black from her head to her feet, except for a small band of white just above her eyebrows; her small face was ducked down demurely on her chest, so that the folds of her veil fell across her face, revealing only a tiny glimmer of chin. Around her neck hung a silver crucifix, and behind her on the donkey was a small rolled pack containing not much of anything.

The boy stalked up to the first gawking peasant woman bold enough to peep out of her doorway and demanded, "¿Dónde está la iglesia, señora?"

The woman, taken aback at the sight of a nun in Pasaquina, quickly curtsied, made the sign of the cross on her chest, and said importantly, "Allá," indicating with her finger the direction of the village square two blocks away.

The boy nodded and stalked on. Behind him, other brown faces peeped out of doorways, brooms pausing in their morning attacks, spoons held aloft, mouths open. As they passed through the square, the old men rose, lifted their hats and made small bows in the nun's direction, but still she kept her head lowered and seemed not to notice.

The boy brought the donkey to a stop in front of the church, let the reins hang in the dirt, and marched off inside, leaving the nun sitting alone on the donkey's back. The talk in the cafés around the square muted somewhat; a vendor quietly pushed his cart of pupusas to the corner of the square nearest the church and stood under a mango tree to watch. But the nun sat immobile, her hands folded neatly through the black strands of donkey hair.

In a little while the boy emerged from the shady gloom of the church, followed by an agitated Rita. She was talking rapidly and gesturing frantically with her hands.

"We did not know," she protested. "The message did not arrive. I swear it by the saints, we did not know. Please, one moment . . . go fetch him . . . he's just there . . ."

But the boy took up the donkey's rope again, planted himself firmly at its head and glowered at her. "No," he replied. "You fetch him."

"But, hijo," Rita protested, "what will he think? He will not believe it if I come. He cannot believe it. If I tell him a nun has come to Pasaquina, he will send me away and say the fairies have got me. No, muchacho, you must go, then he will believe it."

"No."

"Por favor, hijo, he will know I am lying," she pleaded. "No nun has ever come to Pasaquina."

"No." The boy shook his head.

"Oh, may the saints protect us, what am I to do?"

"Do as I tell you, woman. I have a commission from the bishop himself. I am to deliver this bride of Christ to the priest and the priest alone. I cannot leave her until I have fulfilled my obligation to the bishop. I am helping my father."

"Your father?"

"He has been killed in the movement and is suffering in purgatory. This commission is better than a hundred novenas."

"But, child, he will not come with me if I tell him such a thing as this."

"Then drag him by his skirts, woman."

"Ayeiii . . ." Sighing and shrugging, Rita started off down the street in search of the priest, who was only a block down and around the corner, visiting one of the wealthier parishioners. He had gone to bless her house and receive his weekly ration of pork lard, but this process always took some time because after he had made the sign of the cross over the shrine in the corner of the main room and mumbled some words for her, he then had to listen to her complaints about her husband, who drank and whored. She would not relinquish the manteca until he had heard the week's story about her husband's latest adventures. This took about two hours, so he was in the habit of sitting back in a comfortable chair, eating requeson with honey and enjoying himself thoroughly.

Rita had to argue at some length with the dueña's servant before she was finally given permission to enter. And then, as Father Herrera had only in the last moments arrived at the point in the morning's ritual when it was his turn to speak, Rita could only stand in the corner of the room, listening and fidgeting.

"My nephew," Father Herrera pontificated, "is undoubtedly a hero, and, in his own way, no less than a saint."

"Qué grande," sighed Dueña Isabel, imagining the handsome young man in his starched khaki uniform.

"Someday I would not even be surprised if he became a captain in the Guardia Nacional. Why, he has travelled everywhere. No duty is too small for him, for he relishes patriotism as other men love agua ardiente. Nothing is too much to ask of him. No hardship too great for him to bear."

Dueña Isabel smiled piously. "The Guardia is the backbone of the country," she observed.

"Without it, we would be nothing but savages." Father Herrera looked very serious and well satisfied with his own words. "Let the rebels have a foothold and they will destroy everything. They will raise up the greedy campesinos, who can never have enough no matter what the government does for them, and the poor will devour the land like locusts on a corn crop. Soon there will be nothing left for the decent people in El Salvador who built this country."

"Padre, you are right as always," said Dueña Isabel, who thought the priest's statement sounded true, although she was not entirely clear about the connection between the Guardia and the locusts.

Rita could contain herself no longer. "Padre," she gasped, "mi padre, there is a woman, a sister, who has come for you, on the square."

"My child, what is it? What are you saying?" He could see by her face that she was frightened.

"Mi padre, she is from the bishop himself, and the boy will not dislodge himself until he has delivered her to you personally. You must come immediately, Father, please."

"What is this? A woman from the bishop?"

"Tsk, tsk," murmured the irritated dueña.

"A sister, Father, a nun . . ." stammered Rita.

"Oh dear me . . ." muttered Father Herrera. "Are you sure? A nun? Here in Pasaquina? What can she be doing here? How did she come?"

"From the bishop, the boy said."

"What does she want?"

"I don't know, Padre."

"Are there no papers, no writing?"

"No, Padre, at least not for me. You must come. You must come with me now, Father."

"Well, I can hardly believe it, but I suppose we'd better look." He rose regretfully.

"Perdóneme, Señora Isabel, I had better see what is happening in my town."

The old woman laid her arms across her stomach and frowned. "It is best to be about God's work," she said, not knowing what else to say.

"Yes, yes, thank you." Father Herrera backed out

the doorway with a shuffling bow that came from his confusion, entirely forgetting his purpose in coming and leaving the pound of lard, wrapped in its scrap of cheesecloth, sitting on the kitchen table.

The priest and his criada proceeded down the block and around the corner toward the church in a small dust cloud, Father Herrera hustling along in an animated amble, Rita just a half-step behind him, bobbing and chattering and wringing her skirt in her hands. By the time they had made the short trip to their destination, Father Herrera had worked up a good glistening shine of sweat on his rubicund face and was the picture of concerned piety. He came to an abrupt halt directly in front of the donkey's nose and addressed himself to the girl on its back. "Yes, my daughter, I am here," he said, trying to sound important and knowledgeable.

But the sister still said nothing. He was answered by the hard bright eyes and the hard bright voice of the small boy holding the donkey.

"Señor Padre," announced the boy, with more authority in his voice than the priest could hope to imitate, "aquí está Hermana Magdalena. I have brought her to you from the bishop himself!" And with that he handed the baffled priest the hemp rope and marched around to the donkey's rump to begin untying the small pack resting there.

"But, muchacho, for what purpose?" asked the priest, automatically following the boy, and turning the donkey in his wake.

"Señor Padre, I do not know. It is in the papers. The papers sent to you by the messenger from the bishop. The one who came before me."

"No one came. No one at all."

The boy shrugged and firmly turned the donkey back again. "Be still, please," he said. "The donkey turns with you when you turn."

"My child," said Father Herrera, a desperate note in his voice, "I have received no papers from the bishop."

"If you will be still, Señor Padre, I will untie the belongings of this bride of Christ, take a jar of coffee and perhaps some cheese, and be on my way. It is a very long journey. I had to camp four nights on the road. The mountains are difficult and the donkey is slow. If I do not arrive home when predicted my mother will miss me and be sure that I have been captured, so you must be still!"

The priest, dumbfounded, did as he was told. Then, feeling he might look foolish just standing there looking at the donkey, he stepped sideways to face the young woman. She sat as calm and still in the hot dust as a marble replica of herself.

"My dear," stammered the priest, not knowing just how to begin, "my child, can you elucidate this matter for me? The messenger has failed to come. I was not aware of your visit."

Slowly, almost magically, the girl raised her head. The street vendor leaning beneath the shade of the mimosa tree stood up straight and took off his hat with his right hand while with his left, he unconsciously made the sign of the cross.

"Santa Maria!" he whispered. All around the square old men rose to their feet or leaned forward in their chairs, tipping over their refreshments, dropping their handkerchiefs, hissing or sighing or choking out reverent expletives. For even at a distance, they saw what the vendor saw, and the same thought came into all their minds.

The vendor could not say it because it stuck in his lips and closed his throat as with a heavy hand, but it came up into his mouth all the same and he meant to say it, even should it be a blasphemy, which at that moment he thought it was not.

"La Virgen!" he meant to say, would have said if he could, along with all the old men lining the square; for the lovely little oval face that rose from beneath the black woolen veil was nothing less than a sweet, pale cameo of the Virgin, the gentle dove of a Virgin whose quiet black eyes could see all the sorrows in a man's heart and know how desperately he had tried, no matter how miserably he had failed. Eyes that could see and forgive, gazing from beneath long lashes down at the sinner as if from a great distance, and yet with such compassion that forgiveness seemed unquestionable, certain. These were the eyes, this the face not just of a woman, beautiful like a madonna, but of one who *was* the Virgin.

Around the hushed square, the men watched with awe, and perhaps a little fear. What powerful events must be at work that God should send such a one to a place as insignificant as Pasaquina? They saw her lips move as she spoke softly to Father Herrera, too softly to be heard even in the utter silence that had fallen all around her. Father Herrera heard her clearly, however, and his face turned crimson. His head wagged back and forth frantically.

The men were alarmed. Without moving, or so it seemed, they had drawn closer to the tableau in front of the church and could now see even more clearly the amazing sight. There were many miraculous virgins, of course, and all very important, like the Virgin of the Mountains who had her place in the tall trees up beyond the terraces and who was thought to be near to the ear of God and thus very good for helping those gone over. But none was so strong and dear as the Virgin of Guadalupe, whose image was to be seen, on faded postcards and chipped plaster statues, above the flickering candles of corner shrines throughout the village of Pasaquina. There was no doubt in any of the men's minds that here, before them, was the living face of the Virgin of Guadalupe, the Healing Virgin.

But Father Herrera seemed oblivious to this fact. Inexplicably, he was saying, almost shouting, "No, no! Surely there has been some mistake."

The men listened breathlessly, but still could not hear the girl's reply. Father Herrera, however, seemed even more agitated, waving his hands in the air and looking around as if for help. His eyes lit on the boy and he seized the urchin by the shoulders and shook him.

"Tell me what you know of this," he demanded. But the boy just looked stubbornly, perhaps a little scornfully, back at him, and answered in a resolute voice, "She is yours, I tell you."

"But I don't want her!" Father Herrera's voice was shrill now, and so loud that even the few men who had stayed further back could hear him.

"Madre de Dios!" they said almost with one sound. Then the vendor spoke for them all, saying "Take it back, old man."

"You have killed us now, to speak so," said someone else. "God will not tolerate it."

But Father Herrera was so caught up in his own panic that he was completely unaware of the men watching him and had no idea what they were muttering among themselves. His attention was focused on the small boy, who seemed implacably determined to deliver this nun into his keeping and thereby to disturb—in fact, destroy—his whole simple and pleasant life.

"The bishop," said the boy firmly, "has decided that she should be here and has sent me himself to see that she comes to you safely. It is of no matter to me if the messenger before was fool enough to let himself be taken by soldiers and did not bring you this news. I, Chungo, have completed my mission, and that is what matters to me."

A hum of admiration for Chungo's common sense and determination circulated through the attentive audience, and at last Father Herrera seemed to realize that his behavior was receiving their rapt scrutiny.

"But I cannot understand this," he protested in a new tone. "We are a small—a very small—place. What—"

Rita, who had scarcely moved since her first sight of the young nun's face, suddenly dropped a curtsy in the direction of the drowsing donkey, and reached out to tug at the priest's sleeve. He inclined his ear in her direction and listened for a moment, his face gathered up in an expression of mingled consternation and embarrassment, then wiped his forehead with his sleeve. He turned back to the solemn girl on the donkey.

"Forgive me, hermana, forgive me, I forget myself." He actually was in a state of some remorse, which, although it did not stem entirely from his rude treatment of the visitor, did add a note of sincerity to his voice. "Please, step down," he said, offering his hand in what he hoped was a stately manner.

But she remained firmly attached to the donkey. This time she spoke loudly enough that the listening men were overwhelmed and touched by the clear sweetness of her voice.

"I did not know I was unexpected," she said. "I can wait."

"There is no need, I assure you," the priest protested.

"If you wish to put your house in order . . ."

Her face was as calm as a pond with no ripples, and her black eyes made no accusations, but he saw quite clearly that this small girl knew the precise nature of his relationship with Rita, and expected some turmoil within as a result. Suddenly, Father Herrera felt ashamed of what he had taken as a natural and beneficent gift from the heavens sent to satisfy his earthly appetites, even as he satisfied the more ethereal appetites of his parishioners. Shame was not something that Father Herrera was accustomed to feeling; consequently, he had no natural responses, and could only express himself by choking, coughing, wheezing, and finally spitting into the used handkerchief stuck hurriedly under his nose by Rita.

The men around the square saw the priest's behavior as a bad omen, a sign of disrespect. "No good will come of this," they whispered, and crossed themselves for a second time that morning.

"I told you the bishop would not tolerate his priest living with an unclean woman," said one.

"Yes," said another, "he will pay now, even though he is a priest. We should have sent her home after the second son, it is not wise to mix children of different fathers under one roof. I knew God would disapprove."

"La culpa es nuestra. The blame lies with us, for we saw and did nothing."

"Yes, we will all pay . . ."

At last Rita came forward and extended her squat brown hand to the girl, who laid delicate fingers lightly on it and slipped off the donkey's back into the dust. Turning, she followed Rita through the gates and into the courtyard of the church compound. When the last of her black garments passed out of sight, the men around the square, like puppets with their strings cut, wilted back into their chairs.

Father Herrera stared after her in dismay. And the vendor, who was a mountain man and believed his eyes, said just under his breath, "I do not think that it is the bishop who has sent her."

)

*Three* 🌙

Usually eighteen-year-old Guillermo Cecilio Melendez came down from the mountain for his Saturday confession slowly, indolently. Memo, as he was called by everyone in Pasaquina, was almost always dressed in a black cotton shirt open nearly to the waist, a wide black leather belt which held a machete, tight black gabardine trousers with flared bottoms, and black shoes with pointed toes. Most of the other sixteen muchachos, some of whom were a few years older and some a few years younger than Memo, wore baggy khaki trousers, coarsely woven cotton shirts and tattered berets. But Memo, as their leader, felt that he should look different—stronger, more sure of his manhood. And so he tried to appear especially nonchalant and confident whenever he came down to Pasaquina.

This particular Saturday, however, Memo and his muchachos did not come lazily into the village; they came shouting, bounding down the footpath that led across the rocks and through the tall saw-grass to the crooked alleyway that wound in front of the prison, past the post office, and finally into the square. The old men who clustered in small groups there could at first hardly make out the gibberish the muchachos were spouting. Something about fronts changing that made it all sound like serpents slithering through the sand, not at all the proper way to fight a war, and other things about tactics and strategic advantages, which they said meant the rebels were climbing the mountain on the other side.

The old men laughed and said, "Manditos!" The foolish boys often tried to raise these alarms, but the old men paid no attention. No army worth its salt would use Pasaquina's mountain to fight a war. After all, it held nothing to eat but acorns and conejos—or perhaps, they all laughed, the warriors from the left liked to eat flowers. It was a great joke. The only food was on Pasaquina's side of the mountain; on the other side was nothing at all but forest and snakes and creatures too despicable to be mentioned, most notably, vampires.

The older men got the center place then as they described how, years back, their neighbors to the north had lost whole herds of cattle to the vampires, and several children from this very village had even been taken. Yes, they all agreed, it was well known that large, batlike creatures with yellow eyes lived in other forests in other lands, and there was no disputing the fact that such a creature had at one time visited their own forest. In fact, they all admitted, it had never been proved conclusively that the vampire had left. Since it was considered more prudent not to spend the night in the forest on the other side of the mountain, for years no one had ventured over the top after nightfall. Thus, the beast might still be there. There had been rumors.

All of which added up to the simple fact that no soldiers, whatever their political affiliations, would choose the other side of the mountain for the waging of war. It was settled, they decided, and so informed the boys.

"The war will not come to Pasaquina," said the well-respected José Antonio, speaking for his companions. "In the last fifty years this war has never come to Pasaquina, and so why will it come now? No. The matter is settled. The war will not come."

The old men nodded their heads wisely, while the young men fingered the blades of their machetes and looked at one another.

"If you really want something to do," José Antonio said to Memo, "go to the church, for a miracle has passed in this village. A statue has come to life. The Virgin has come to rest among us, and if you would see magic, then all you must do is go to the church and take one look at her face. Then you will see something to make you forget the war. You will see something that is real and that has happened in our village, not just the spoutings of young hotbloods."

The young men spat in the dust and cried at their elders with great fervor. "You're fools!" said one of the muchachos, his voice thin with fatigue. "Look at you, sitting here placidly like sheep, starving, and we have no schools for our children and no medicine for our old ones in this, the modern age!"

"It is the Church itself that keeps you asleep, with its fine opium fumes of high ideas and false promises." Memo spoke now, harshly. "I don't know why we haven't torn it down already! If we were men, we would burn it to the ground as a service to the people."

"Then you couldn't go to confession," giggled an old toothless one, his face like a mummified skull.

"To hell with confession!" shouted Marcos, a young man of twenty-one who had never found work and had eight brothers and sisters at home with his widowed mother. "To hell with the priest!"

"Then why do you go to confesson so faithfully each Saturday?" taunted the old man.

"Hijo de puta!" Marcos shouted, turning on his heel. "I will burn it down. You wait! It holds you slaves, all of you!"

Memo gathered the muchachos with his eyes and they moved haughtily away from the table of the old men. Only one young man, who had been on the mountain just two months, lingered for a moment.

"It is for his mother he goes," he told the café cronies apolegetically. "She said she could not bear to lose his body to the war unless she knew his soul would be in heaven waiting for her. She made him promise."

The old men laughed derisively. "For his mother, is it? No, man. Maybe today, but let him take his first bullet; let him see his own blood spilling in the dirt, and then he will go for himself, like the rest of us. He will go for his life then, not for his mother." José Antonio spoke eloquently, his companions thought, and they nodded sagely as they rocked in their chairs.

In the meantime, Memo was making his way past the promenade, past the dry fountain in the center of the square to the cantina on the opposite corner. The others clustered around him, moving in his wake like bobbing boats, waiting for instructions.

"We need recruits," Memo told them sharply, flinging the words over his shoulder. "And we'll be needing food, lots of food. When they come from the other side of the mountain, we must be ready to supply them. We've got work to do."

"Would Marcos really burn down the church, Memo?" one of the youngest inquired in an awed tone.

"Sí, he would. I would myself. We're fighting for freedom, don't you know! One form of slavery is no better than any other."

"Then why haven't we done it before now?" asked Arturo, eager for action and for some sign that they belonged to a larger group than themselves.

"Because it is our village. The people would not understand. They would give us nothing but trouble. They might even report us to the Guardia. Who knows?" Memo shrugged. "It wouldn't be worth the effort."

"Maybe the people would join us," said Marcos, who had returned to the group after a fuming walk to the edge of the village and back. "Maybe they would see that we mean business, and they would listen to us."

Memo shook his head impatiently. "They are lazy fools. They will do nothing but stare."

"But," Marcos insisted, "in many places the people are with the revolution. I know it is so. They help the guerrillas."

"Not in Pasaquina," Memo said grimly. "Not now, at least. When the revolutionaries come up the mountain, when the others come, and they can see the soldiers and see that the war is real and that the boys are fighting and giving their lives for the freedom of the people, then they will understand."

"Will we burn the church then?" begged Arturo, his face flushed already by imagined scenes of the burning.

"If the Father holds out on us, we'll see," muttered Memo. "If he doesn't give us food and shelter, if he doesn't take our side, then we'll see."

"But it is a church, a house of God," Eduardo, who was barely fifteen, suggested timidly.

"It is a dung hill full of politicians who support Las Catorces Grandes," Memo said exasperated. "Don't you know anything, fool?"

"But—it is only Father Herrera." The youth was still befuddled.

"Don't ask me any more stupid questions, hijo. Just look at the size of his belly, and then look at your own if you want answers."

"Sí," sighed the boy, not understanding entirely but accepting and still thinking about the bonfire. He wondered what would happen if the crucifix over the altar caught fire. He thought most certainly the mountain would split in two and bury them all alive at that very moment. It was the most dramatic thing he had ever envisioned, and he felt he would like to see such a thing, even if it would be the last thing he saw.

"Make the rounds, muchachos," ordered Memo, as his back faded into the darkness of the cantina doorway. "Get what you can. I'll find you at sunset."

The boys started off, not very optimistically. There was not one household in Pasaquina with food to throw away. There was not one household that believed that the war was real, and they themselves, though among the chosen, were not always entirely so sure.

Throughout the afternoon, sixteen of the young men—the majority of the band—scoured the town, begging, rolling their eyes soulfully, and invoking the spirits of all the saints they could remember, the departed loved ones of whomever they were speaking to, and in cases of dire necessity, even the local demonios. In return, they received a few cold tortillas, half a dozen eggs, a skin of goat's milk, one thin young rooster, three handfuls of cacahuates, eight half-rotted bananas, a wilted cabbage, and the priest's forgotten pound of manteca, which the dueña gave them out of spite. No beans, no coffee, no rice, no corn, and no meat, all of which they would need to feed extra men. Finally, just at dusk, they wandered back to the square, sharing home-rolled cigarettes and shrugging their shoulders laconically.

"They will never come anyway," Arturo said to dismiss the trouble.

Then they all felt at peace with the evening, no breeze yet but the dust settling over the square and the scent of cooler air coming down from the mountain, until Eduardo, who didn't know enough to keep his mouth shut, said: "If they do come they will surely be very rich and will bring plenty of food of their own. They will be so rich they will feed the whole village. Then what a party we'll have!"

The others glared at him from under their berets until he blushed and was forced to yank a cigarette from his nearest comrade and attempt to take a long drag. He hid his face when he coughed the smoke out again. But the boys wouldn't speak to him again that night.

When Memo came out of the cantina he was drunk, really very drunk, drunker than they had ever seen him. Memo was their leader not by reason of age or seniority, but because he had the darkest, fiercest complexion of them all, the hardest scowl, and the fastest, meanest hands. Memo had cultivated this appearance ever since he had found that he could not bring himself to take up his dead father's plot of land on the terraced mountainside and spend his days bending over a hoe. He burned inside with a need to do something great. So he had left his plot in the hands of the smaller children of the family and joined the men of the mountaintop to wait for the war to come to Pasaquina.

At first there had been on the mountain only a small gathering of ne'er-do-wells, thieves, wanderers and the like, but then their numbers grew, as if the burning desire Memo felt was a contagion that wandered off the mountaintop on the night airs and infected more and more young men until they could no longer resist its pull, and trudged up past the terraced fields, following their vague visions or their muted inner voices which never spoke clearly, so full of desire, so distended with longing that they were near to explosion by the time they reached the top.

During the days, they slept or foraged for maranones or pepetos; occasionally one of the older ones would take an expedition south toward the highway to waylay norteamericano tourists. But whatever they did, the days were only a form of waiting. They waited for war, and the better life that was sure to follow, and for the nights, which were full of mystery and magic. It was then that the older ones sat around the campfire, sometimes drinking from a common bottle, sometimes with the luxury of cigarettes, sometimes chewing on a twig, and told the stories.

They told about the heros of the past, saying, "And at the very moment they stood Dalton up against the wall, and raised their rifles to shoot him in the eyes, the earth began to shake and tremble until it fell from beneath their feet. The buildings fell on their heads and crushed them and turned their rifles to splinters, and the whole nation shook from Ahuachapan to La Union, so Roque Dalton could be free. God sent the earthquake. That much is clear. You can see by the situation that God is on our side."

"Es cierto," the younger boys nodded in agreement, which encouraged the older ones to pursue their campfire reflections.

"Of course God is helping us, after all, this is the only piece of heaven left on this earth—does not its name tell you so? It is all that is left of our far home, the only place left where the fine oriental soul of the Maya entwines itself with God's clay bodies. We are the last of his children. Naturally, he will protect us."

These were the things which filled the boys' heads as they waited for Memo on the square that night. But they were hungry and cold and very tired and the heros were so long dead and the places of their dying so far away and strange, which is why, when they saw Memo reeling toward them, they felt a twinge of uncertainty that was not entirely due to his inebriation or the fierce yellow glint in his eyes. If the rebels really came, it would make a difference. No one could think just what kind, but they knew it would.

Memo said nothing about their lack of success. He only looked at them viciously, spat in the dust and headed toward the edge of the village. Though he stumbled from time to time, he did not weave when leading his men.

The silence was oppressive, and finally Eduardo could stand it no longer.

"But, Memo, we'll be back on Saturday for confession. They'll give us more then. You'll see, they'll be more generous on Saturday."

"No!" snapped Memo. "No!"

The boy was troubled and looked to a companion for guidance, but his comrade only shook his head.

"No," continued Memo, to himself it seemed, "we will not be back for confession on Saturday. We will never be back for confession. I have been lax and a fool to let you come. It softens you. It softens us all. And it is against the revolution. You may not come back. I may not come back. Never again may we come back to that pit of vipers belonging to the fat priest."

"But it is just Father Herrera . . ." someone ventured.

"Sssssh," hissed Arturo, "he's only drunk. Wait for tomorrow. See what he'll say in the morning."

They continued up the mountain trail as quietly as small brown mice. As he walked along, Eduardo wondered if Roque Dalton had been as fond of chicha as Memo was, and whether all the rebels, if they came, would be heros, or whether some would be like himself.

)

Four

In the village, the muchachos were forgotten before they had even left. Nothing was of interest to the town now but the presence of the Virgin, and how long it would take before she removed the unclean Rita out of the church forever. But Rita, having felt herself somewhat in jeopardy ever since the birth of her second child, who was growing to look more and more like his real father every day, had taken the young nun, Sister Magdalena, into her personal custody. She had scoured out a room Father Herrera liked to call his study, which was stacked with yellowing newspapers, seldom-read books, and copies of letters he should have answered or burned years ago. Rita took the entire mess, load by load, and dumped it, without even bothering to ask permission, into an unused corner of the padre's bedroom.

The newly emptied room was large and windowless, with primitive double doors made of thick oak and fastened shut by a four-inch oaken beam that passed through iron hooks in the middle of each door. It had once been used as a private chapel for the priest, the original priest whom no one could remember, and had one wall of faded yellow, one of pink, a third of white (or what had once been white), and on the long back wall, a painted mural depicting the village at sunset. In the mural, Pasaquina was shown with numerous green trees it did not possess, a splashing fountain, clean, happy children, and in the center, the church, looking as it had never looked in reality. There were even tiny doves circling around the small bell tower, though if the picture were to be believed, the bell must have been ringing loudly at the time. It was easy to see that the artist had enjoyed himself, had indulged himself with color of a kind unseen in Pasaquina, even in the springtime. But the colors were faded now, and chipped in places, and the poor, empty room had a forlorn quality, something like the essence of a child's funeral.

Still, Rita beamed proudly at the results of her efforts.

"It's a fine room, hermana," she exclaimed. "Once it was a grand room, the grandest in the whole town, and used only for a chapel. No one has ever slept here, only prayed. That makes it especially good."

"Thank you," replied Sister Magdalena. "It is very lovely."

"I will find everything you need." Rita's small eyes danced with excitement, for she had momentarily forgotten her position and felt more like she was helping a friend settle in.

"I have everything I need here," said the girl, indicating a small pack on the floor.

"No. No. Absolutely no," said Rita, and she was firm beyond her station.

So she provided the girl with a linoleum-topped table, a straight-backed chair, and a long bench, which she set in front of the mural and covered with an embroidered manta cloth; on it she placed a candle in a saucer and an empty vino bottle which she had filled with wildflowers. "This can be the altar for your prayers," she explained happily. The girl nodded, and after Rita had bustled out, smiled a young girl's smile in secret.

After much activity, many loud exhortations, and even the enlistment of Father Herrera's unwilling bulk, Rita produced a bed of sorts by stacking red clay bricks, spreading three boards across them, and dragging in a very thin, much used cotton mattress. "A donation!" she explained proudly. Linens were no problem at all; Rita simply sent Father Herrera begging to Dueña Isabel. Grumbling inwardly, the priest nevertheless bartered prayers for sheets and a cheerful patchwork quilt, and trudged home again, thinking it had not been a particularly pleasant day and hoping Rita would stop soon and cook something special to make up for all his difficulties.

When the room was finished at last, however, Rita demanded that he have a bath before dinner—an amazing and distasteful idea. But he was too tired to argue, and so went away with a weak smile on his face, feeling a very beaten man.

"It is nice," said Sister Magdalena, when Father Herrera had left. "Thank you."

"It is nice," agreed Rita. "You like it?"

"I will be very happy here, I'm sure," replied the girl, keeping her face as still as morning air.

Rita paused on the threshold, curtsied, and then said, "Please, hermana, forgive me for saying it, but you are very beautiful. The whole town is saying it. They are saying so many things. They do not believe you are real. They believe you are the Virgen de Guadalupe stepped down off her pedestal and come to life. I have been with you all day, and I, myself, also almost believe it too." Then she backed up hurriedly, frightened by her own temerity. "I'm sorry."

The young nun's large black eyes stared at her without expression for a moment, and then the edges of her lips began to quiver. All at once she began shaking and covered her face with her hands. Rita, thinking the girl was crying, rushed to her and threw her arms around her shoulders.

"Oh, no . . ." she wailed, always quick to cry with a sister. "No, no . . . look what I've done."

But in a short while the shoulders Rita embraced stopped shaking and the girl peeped out from behind her fingers. There were tears in her eyes, but they were tears of laughter. She had been giggling. Seeing Rita and her mournful face, the laughter rose up in her again and she broke out into another spasm of giggles.

"Oh, dear," exclaimed Rita, for all at once the Virgin had been transformed into something very like her own little sister. "Oh dear, oh dear," she repeated, and then she was laughing too, still holding the nun gently by the arms.

Finally, Sister Magdalena caught her breath. "I'm sorry," she whispered, with an impish grin on her face. "I'm so terribly sorry. Here they are, all thinking a miracle has happened and I'm laughing at them. It's so horrible of me. The Mother Superior would snatch my hair out for that!"

"¡Qué terrible!"

"She would, truly," said Sister Magdalena. "You've no idea how mean she is. She hates us all, believe me. What she wanted was to heal lepers in Africa and what she got was a convent full of silly girls. You can't blame her really, but it's terrible to see her angry." The girl looked troubled for a moment, and then her face relaxed into the look of one who is tired.

"I'm sorry. It's been such a long trip, perhaps that is why I forgot myself. Please forgive me, Rita."

"No. No. It is you who should forgive me, lady. I am the one. I should never have repeated the gossip to you, but it was so like the truth. I needed to be sure, you understand. It could have been the truth. I wanted to check, just for myself, that my eyes did not deceive me, you understand?"

Sister Magdalena's laughter broke out all over again. "I can't help it," she gasped. "It's so funny. It's so terribly funny."

"Please?"

"Will you keep a secret, Rita? Will you be my friend?"

"My word of honor," vowed Rita, suddenly solemn again.

"It's so terribly funny, you see, because I never wanted to be a nun in the first place."

"No!"

"Yes, it's the truth. I'm only here because I couldn't do anything else."

"Oh no, this is terrible! And they think you are a saint!" Nevertheless, Rita giggled again behind her cupped palm. "But if you did not want to be a nun . . ."

Rita's words were stopped by the look of sadness on the other girl's face.

"I wanted only to be like any girl," Sister Magdalena

said softly. "To grow up and have a novio and then be married in a white lace dress and have many happy babies. I thought always of this, even when I was very little— thinking of my husband and children. Even when I was four, though you will not believe it."

In truth, Rita found it hard to imagine, although she wrinkled her brow and tried very hard. She herself had never dreamed of anything in particular and had been glad to find such interesting opportunities in life as had come her way. Briefly, she thought of Manuel, then shrugged and turned her mind back to the sister, who was surely crying this time.

"Ah, hija, you must tell me what has happened to this wish," Rita said to the sobbing girl. "Come, sit here on the bed and tell me."

The young nun allowed herself to be drawn down onto the newly constructed bed, letting Rita's arm encircle her shoulders comfortingly. But then she shook her head, sprinkling teardrops so that they splashed on Rita's own cheek.

"I cannot explain," Sister Magdalena said with a sigh. "It does not matter anyway."

"But of course it does," exclaimed Rita, who wanted very much to find out this mysterious story. And, too, what if Sister Magdalena were to cry all the time? The thought of this beautiful creature perpetually unhappy made Rita feel sad herself. And, she thought practically, it could not have a very good effect on the household, to have such an unhappy one about. "You must tell me, little one, you must!"

Sister Magdalena raised her eyes to meet Rita's. This

woman was not all that much older than herself, she marveled, and yet so different. How to explain to one who knew only the simple life of Pasaquina how cruel the world could be? But then, it would be so good to speak to someone of the pain she carried with her.

"When you have told me," Rita said invitingly, "then it will all be out of you and we will feed you and tuck you into bed and you will be happy again. Why do you not marry if that is all you have wanted for this long time?"

After a little silence, Sister Magdalena said softly, "I have been in the convent since the age of eight. There one does not meet the young men. Soon it is only possible to become like the others, a nun."

Rita could see that this would be true. "But was there no other place for you?"

"No, nowhere. My family is all gone, you see, all killed."

"All?" Rita sucked in her breath. "Killed? But what happened?"

"They were all shot, again and again, shot all around me, by the Guardia." Now that she was saying these things, Sister Magdalena found that it was easy and felt pleasurable. She wanted to say more, to listen to the sound of her own life being told like a story.

"But, no!" cried Rita. She had never before known anyone who had known anyone who was killed on purpose, like an animal being slaughtered. To kill someone in a fight, yes, because the passion and the anger spilled over and then someone was dead. But a war was not, in Rita's opinion, like a fight; it was not personal and important like the anger of two men over a woman or an insult to one's honor.

The other girl looked far away, her face very still and thoughtful. Finally, she spoke. "Oh, it is true. Such things may not happen here, but in the city, and elsewhere too, it is often true that people are killed, and their whole families, because they oppose the government. It is even so that those who are shot are fortunate, for many are not killed with clean bullets but are hacked to pieces by the long machetes of the Guardia. Some have their faces taken off inch by inch while they are yet alive."

Rita's face was frozen in an expression of amazement. That the sister could speak of such things so calmly —for now the girl seemed almost peaceful and her voice was cool—was startling enough; that such practices truly happened was almost beyond belief.

"My father was one who believed something could be done against the government," Sister Magdalena was continuing. "He was trained in the law, but he said that there was no use in it anymore, and so he opened a business. Soon, because he was very smart, the business had grown, and he had to buy the shops on either side. We were not rich, but we had more than most. We were never hungry."

Rita understood this clearly, for now that she was with the priest, she herself had more than most people in Pasaquina. The idea of having even more, a shop the size of three shops and all the things in it and all the money, was intriguing.

"But," the young nun continued, her voice still serene, "my father could not be content while others, the laborers and the farmers, had so little and were always held down by the government. He and some men he knew believed that if the trabajadores had leadership, perhaps a revolution could succeed and the government could be overcome."

"I see," Rita said carefully, to let the sister know that she was listening and thinking about the story.

"So the men planned and they schemed and they gathered together all that were left from the area to take the barracks of the Guardia." Sister Magdalena's face was growing more pale as she talked. "But there was a traitor among them. The Guardia was informed. They laid a trap. From in front and behind and on both sides, they shot at the men with canons, and rifles and machine guns."

"¡Qué terrible!" exclaimed Rita. "And your father was there?"

"Yes, he was there. But by some miracle, he got away. He must have crawled beneath their boots, between their legs, in the height of their killing lust. It is the only way he could have escaped, don't you think?"

Rita nodded, not knowing what else to do.

"He came home to my mother and he was hurt very badly, bleeding everywhere. He could not live till morning, even I could see that. But he came home to my mother to die. Somehow, someone must have seen him, or else they followed the trail of blood. I've never known how, but they found him, and to make examples of us all they began firing into the house, three of them abreast, with machine guns."

The level voice went on. "It is the sound I remember most. The noise of those bullets tearing into the walls, and above it all the sounds of my little sisters screaming. I even heard my mother. She was trying to pray to God to save us, her babies, but she was hit in the throat by a bullet. She never finished the prayer. I wondered, while it was happening, how I could hear so much—the screaming, the broken prayer—how could I hear it above the awful sound of those guns? I can hear them even now. They hurt my ears they were so loud, so terribly loud."

"But pobrecita," asked Rita, in awe, "how are you here?"

"I? I fell. I fell from the bed in which I was sleeping with my sister, and in fear crawled beneath it. It was a low bed. It happened so quickly, they could not have seen me. Only that, that it was all over in a moment, and the fact that it was a low bed, built for children, only these little accidents saved me."

"And they were killed, all of them?"

"Yes. I think so . . . no." She shook her head, as if at a hopelessly difficult puzzle. "I don't know . . . there was my brother."

"Your brother?"

"My older brother, Roberto." She said the name like a kiss. "He heard them coming. 'Run,' my father said, but we could not. There was no time. As I fell I saw Roberto go toward the back window. Then a barrage of bullets hit the window and he was gone. I do not know if they killed him. I have always hoped . . . I have always prayed for Roberto."

"You didn't find him? You didn't look for him?"

"I could only lie beneath the bed when the soldiers had gone, I could only lie there and could not move. The fear was so great it made me soil my underthings, and then I was so ashamed I could not come out."

"Pobrecita . . ." murmured Rita.

"I lay there for two days, and then a neighbor

woman came. She heard the Guardia was going to take it all, the store, the house, everything. She came to look and see if it was true that we had been all killed. You understand, no one came out. No one looked outside his door in that block after the shooting. She was the first. When she found us, she cried so hard it made me cry again too, only louder, and that is how she found me, beneath the low bed of my sister."

"Pobre, pobre, pobre . . ." sighed Rita.

"She hid me and went back to the Father. I was taken to the convent in the middle of the night. I was dressed as a boy, and it was the same night they came to take away the bodies of my family. They threw them in a pit, stinking, covered with flies, three days old in the heat. They shoveled them like garbage into a wagon and threw them into a pit without a marker to say their names. But I do not know if I saw that or only think I remember seeing it because it was told to me later."

"It is a terrible story," said Rita and then stopped, uncertain of what to say next. It was important to say the right thing, so that the sister might feel better.

"But now you are eighteen," she said at last. "It is all gone. It is all over. It will never happen again, that much you can be sure."

I know it is over," replied the girl softly. "But somehow it still seems to be here, as if it were always happening, again and again in my head. That is strange, do you not think? That it should happen to me again and again like that, for it is certainly over a long time now. I know that is true."

"And it is also far away," insisted Rita. "It did not

happen here. The war will never come to Pasaquina, believe me, it will never come here."

"I believe you," said the other girl. "No one could ever make a war in such a small place as this."

And then she smiled at Rita.

"You are better now, hija?" Rita asked her.

"Yes."

"Then I will go cook something nice for you. You will feel better after something nice to eat."

Sister Magdalena's face lightened. "We never had anything good to eat in the convent."

Rita laughed. "You will always have something good to eat here." But she turned back just as she was about to leave and asked, "What is your real name, then, your name before you became Sister Magdalena?"

"My real name is Rosa, but my father always called me Luna, because he said my face was white like the moon when it is full."

"Luna. Luna with the white skin," Rita murmured. "That is beautiful, just like you. Your father was a poet. And Roberto? Did he also have light skin like the moon?"

Suddenly the oval face was still again, dropping out of animation as if a mask had been pressed over it. "My brother Roberto was a student, but—" she faltered, "I don't know . . . I can't . . . I can't remember his face."

She gestured helplessly, her pale, small hands fluttering. "Sometimes in my dreams I see him, but when I wake, he is gone, and I cannot remember how he looked. No matter how hard I try, I cannot see his face." She paused and bowed her head. "It is very strange. You will think me odd."

44

"I think only that you are tired and hungry, and you have talked of painful things. I will bring your supper to your room on a tray tonight. Would you like that?"

"Thank you, yes. If you think the Father will not mind."

"Do not worry about the Father," Rita replied crisply. "I will take care of the Father." And with that, she vanished through the door.

Alone in her chapel room, Sister Magdalena looked out through the double doors into the growing dark. Down the open corridor she could hear Rita rattling pots in the kitchen and scolding the priest. Far away, children were laughing in one last frolic before bedtime.

She lay down on her simple bed, closed her eyes and saw in her mind's eye the faces of her father and her mother and her baby sister, but she could not remember the face of Roberto. *It must have been a beautiful face,* she thought.

*Five*  )

By the time Memo and his scraggly band had regained the mountaintop, the moon had passed behind the clouds and it was dark even in the small clearing that served as the muchachos' camp. The clearing backed itself up to the mountainside beneath a protective outcropping of rocks, under which a small cave had been dug from a natural fault. Near the opening of the cave, a stealthy light from the low-burning campfire barely eased the night's blackness. As the flames wavered, they cast an uneven glow which moved over the drawn-up knees and hunched shoulders of the men squatted around the fire, a brief light passing across one bearded face, then disappearing, next catching the side of a lean jaw, and quickly flickering on to the next or back again.

So it was not until Memo and the others had fully exposed themselves, coming close to the oasis of firelight and dropping their burdens in somewhat exaggerated exhaustion, that they saw the newcomer seated next to Ramiro.

"¿Qué tal?, Memo," the grizzled old Ramiro called. Ramiro was a gypsy, over sixty years of age, and not a part of Memo's band, though he guarded their camp with an old rifle and fished in the stream for guapates and pepescas, which he cooked over the open fire for the boys. Sometimes he shot squirrels or wild boar, but that was rare.

Memo pulled himself up sternly erect and stared at the stranger beside Ramiro.

"Un amigo has joined us," Ramiro said cheerily. He was far too drunk on the stranger's chicha to worry about formalities and security.

Memo didn't answer, and the stranger made no move to rise, so Ramiro rumbled on hoarsely.

"He is a comrade, a guerrilla—very fierce. There is no one more fierce!"

"¡Cállate!" snapped Memo, glaring at the stranger. Behind him the small band of young men stood, legs apart, their hands on their machetes, uncertain of what was happening, but ready to follow Memo's example.

"¿Qué pasa?" asked Memo, standing absolutely still. The stranger was completely still as well. Nothing about him moved; there was no sign of restlessness, not even the batting of an eyelid.

"Nada más," replied the stranger.

Memo flicked his eyes at Ramiro but no explanation rested in the old man's face, so he continued with as much bravado as he could muster, for by now he had noticed the stranger's high-topped boots of good quality black leather, his dappled army fatigues, and the shiny black automatic rifle across his knees. This was a well-equipped man, and from the stillness of his frame and the creases on his face, Memo saw he was a man accustomed to combat.

"How did you come here?" demanded Memo, meaning, why did you come here.

"To see you," answered the stranger.

"¿Cómo se llama?" asked Memo, still inhospitably.

He gauged from the stranger's bent frame that, standing, the man would be no taller than himself, if as tall. Memo thought he would be not more than five ten or eleven, but he saw that the arms extending beneath the rolled-up sleeves of the man's bush jacket were tight and sinewy. He guessed that the rest of him would be the same, stringy and taut, but with a strength to be respected.

"Beto," replied the stranger, his black eyes glittering in the firelight, glittering and evil like the eyes of a witch's cat, Memo thought.

"¿Y tú, amigo?" the stranger asked, using the familiar form; whether it was used to insult or to woo him, Memo could not tell. From any other man, this sudden familiarity would have been intolerable, but Memo found himself answering the stranger's question automatically.

"Guillermo."

The old one rumbled gleefully, waving the calabaza toward Memo. "Yes, this is our Guillermo, this is our leader, a great man. A ferocious warrior, a wild boar. When they see him come into the village, the women faint and the old men tremble. He can wither a flower on its stalk with one glance." Then he giggled and drank more chicha.

"What do you want?" asked Memo, looking hard at the stranger.

"Sit down comrade," the man said gently. "Let us get to know one another. Have a sip of my chicha. It is exceedingly good, I promise you, only the best."

But Memo would not budge, though he felt himself almost beguiled by those small, black eyes. Already he was stirred by the man's power. He wanted to listen to the smooth voice and drink the chicha, to hear stories from the other side of the mountain, as one soldier might listen to the

48

tales of another. But because he wanted to be flattered and praised and accepted by this man whose power he had begun to sense, instead he stood his ground resolutely and scowled beneath his thick brows.

"What do you want?" Memo repeated.

"Come, let us get acquainted," insisted the stranger, indicating a place on the ground by the fire. "Perhaps we shall spend a great deal of time together in these next few days. It is better if we get to know each other."

Memo resisted. He yawned, spread his arms in the air to stretch himself and said, "Dios mío, I am tired. It had better not take long. I have many important things to do in the morning."

Slowly, as if it were only because he was tired, Memo moved toward the fire and stretched himself lazily at full length opposite the man.

"Where do you come from?" Memo inquired, feigning indifference.

"Oh, over there," answered the stranger.

"You have come a long way?" asked Memo.

"Not so far," replied the stranger. "And yourself, where do you come from, Memo?"

Memo bristled at this further familiarity, but attempted not to let it show. "Many places, oh, many places. I have been everywhere, camarada. Everywhere in this world and the next." Then he laughed boisterously at his joke. The boys laughed too, but softly.

"Yes, I have heard you are an experienced man," said the stranger who called himself Beto. "A wise man, even, I have been told. Is that true?"

The flattery was beginning. Memo knew it. He hated it. To accept it would show his weakness and show the stranger what kind of man he was. But the feel of it was like the caress of the woman he had been waiting for all his life. The boys, too, listened with passionate attention, as if they had been waiting for this music a long time. They wanted Memo to let the stranger roll on with his words and spread out his flatteries until their bodies were massaged with the words like sweet oil, until the contentment satiated them, until their dreams of noble war, of courage, of daring, were no longer dreams but the truth because the stranger said it.

Memo could not resist for the same reasons the boys did not wish him to, and so, lying on his side by the campfire, he let his eyes rise to Beto's and allowed a slow, broken-toothed grin to spread across his warrior's face. Now, in Beto's presence, he felt himself a warrior.

"No, no," he said softly, good naturedly. "It is all lies. I am as wet as a child in his mother's womb. I am as pink as a gringo. No, man, not me. I am as delicate as a girl. If you heard otherwise, you've heard nothing but lies."

The stranger laughed. It was a good joke between the two seasoned veterans. "Well then, I have certainly heard many lies about you, Memo. Tell me, what shall I do? Here I am, come to see you all this distance because I have heard of your strength, of your intelligence, of your superior courage, and now I find you at last, only to discover it was all lies! What am I to do, Memo?" He laughed again. Memo laughed with him.

"Could you change, Memo?" the stranger begged, with a teasing grin around his thin lips. "Could you change yourself so that I would not have wasted my trip?"

His face flushed with pleasure, Memo reached toward Ramiro for the gourd of chicha. He took a swallow, rolled it on his tongue, and found it just as good as the stranger, Beto, had promised. "Tell me," he asked mockingly, "how many of these jugs of courage have you? Enough for all my men? They are all nanny goats and suckling pigs, not a man among them. They could all use some of your sweet courage."

Beto laughed. "I think I have enough." He picked out a boy with his eyes and indicated a pack laid outside the mouth of their tiny cave. "Everybody have some. Sure. Everybody take a drink."

Memo glowed. He was still a hero.

The boy took only one jug from the stranger's pack, though there seemed to be a month's supply, and tiptoed quietly back toward the others. As the boy came behind the stranger, Beto reached out and grabbed the calabaza from him. He pulled the cork out with his teeth, spat it onto the ground and raised it in front of the firelight. "Here, let us drink to you, Memo, and your boys. I do not believe one word you have told me. I believe they are fine men, all of them, and courageous soldiers, every one. And I believe you, Memo, are as quick with your inventions as you are with your hands. That is very good, Memo. I salute you." He upended the bottle and let the brown liquid spill out the corners of his mouth.

Memo did the same, drinking lustily, the eagerness overtaking him. The stranger handed the bottle back to the boy and leaned on his elbow, smiling at Memo. "I am pleased with you," he seemed to be saying. Memo could almost hear him say it: "You are far better than I expected." The stranger's small, wistful smile beamed the thought onto Memo's face, and it felt like sunshine on his cheeks. So he had not after all lost face by accepting the stranger's praise. He had not shown himself to be a foolish man, unworthy.

He was full of contentment, ready to be generous with himself.

"Who are you really?" he asked the stranger playfully.

"Like you, Memo. The same as you, no different" was the reply.

Memo laughed and drank again. He knew the stranger *was* different, but he liked the sound of it too much and let his heart be convinced so that the sound would continue.

"I am like you," Beto continued, beginning the spell. His audience sat rapt and attentive, ready for the hypnotism, pining after the trance. "I am a man of the people. A simple man who loves his country. I am a revolutionary who fights for the freedom of the people, no different than you, Memo."

Through his alcoholic haze, Memo suspected the stranger knew that he, Memo, had never fired a gun at a living target in his life. But he loved the words. "Tell me more of yourself," he suggested.

The stranger laughed. "It is not a good story," he said. "I have only humble origins. There is no story at all," he assured them, "until I joined the revolution, and since then I have been like you, only a homeless, wandering soldier. So, you see, you know the whole story already. It is so like yours, there can be no difference worth mentioning."

"Yes," nodded Memo.

"I have seen a few skirmishes," the stranger admitted, "but they are nothing like yours." And so the legend became reality in the stranger's mouth, and Memo at last had become a fighter. "Tell me," said the stranger. "You tell me. I would like to hear of your adventures. Anything I could say would certainly grow pale in comparison." He laughed softly. "You see, you cannot fool me, Memo. I have already heard about you."

Memo grinned lazily. "Then if you know, why must I tell you again?"

"I would like to hear for myself," said Beto. "To know the truth. And besides, they are very good stories. You must entertain your guest, Memo," he teased. By now, Memo no longer felt offense at the sound of his pet name in this man's mouth. It had become an intimacy he cherished.

He yawned. The boys grew still. Ramiro snored. Beto sat quietly waiting; nothing about him moved. Memo thought it strange that a man could sit so still and decided the stranger had no lice.

"All right," he said, "for my guest," and took another drink. "I left my mother's skirts when I was only a lad." The stranger smiled approvingly, so Memo forgot him and wandered on with his tale. "I was only thirteen, a small boy. I did not grow to this size until much later. I will never forget how she wept, but I told her 'No, Mama, I must go, do not try to stop me. I am a man, and I must defend my people. If we will be free I must fight, for I am a man.' " Memo seemed to get lost in the thought, so the stranger encouraged him.

"You are a man. Un hombre grande."

"I travelled alone through the mountains, never stopping at a village for fear I would be caught and sent home again, due to my small size and extreme youth," Memo continued. The boys listened intently. It was not the story they knew to be true, and consequently, it was exciting. Memo was inventing it one sentence at a time, they thought, but with the stranger from the other side of the mountain present, perhaps it could be true.

"At last I had covered fifty kilometers without one stop and nothing to eat but a few tortillas and an onion. We did not have much in our house. There was nothing to spare. I do not know how I found them, but I think surely something must have guided my footsteps. I was almost dead, I can assure you, before I discovered their camp." Here the narrative grew sketchy, for he had not travelled and could not visualize any other camp than their own. He took a drink to give himself a pause for thinking and then pushed ahead.

"They took me in immediately for they could see what a brave lad I was. They told me after I had been fed and had slept a while that they would send me back with an escort, and that my journey would long be remembered and serve as an inspiration to all the men. If a boy could show such courage, then what right had they to faint from battle? So you see, I was a great success in the camp. They treated me like their pet and gave me the best of everything. Roast meat, quail, raisins, oranges, everything you could want."

The stranger did not smile, though he wanted to. Raisins, he thought, when have I seen a raisin in my life? When does a band of rebels march on quail? Quail we sell. For ammunition. We do not eat quail. But he nodded gravely at Memo, biding his time, doing his job as he had been trained to do at the beginning of his lifetime.

"They wanted to reward me for my courage and so they treated me like their pet. Then they said they would

54

send me home a hero. But I would not hear of it, not me, Memo, a boy of exceedingly small frame who had walked fifty kilometers through dangerous territory alone. I had not walked so far to be sent home again. I told them so. At first they tried to bully me, and said they would throw me out of the camp if I would not go home. But I would not budge. 'Proceed!' I declared. 'I will only camp in the forest by your side. When you move, I shall move. I shall follow you to the ends of the earth, for it is my greatest wish to fight for my people.'

"Then they begged me. 'Memo,' they pleaded, 'can we not send you home again to grow some? When you are taller you can come back again. Then you can fight. You can even lead us. We promise you.' But I would not hear of it. They didn't even tempt me. Not for one moment did I waver in my course. 'I, Memo, will fight now,' I told them. 'You cannot dissuade me.' "

The story rolled on now. Memo heard himself speaking with such confidence that he wondered if he might have had another life and was just now remembering it.

"At last they had to accept me," he continued. "What else could they do? But they were exceedingly proud of me. I was their mascot. But I fought with them," here he faltered again, not ever having actually seen a fight, "fought with the best of them, so many dear comrades, so many noble men. Oh, how brave they were. Fearless, every one of them."

"To be sure," nodded the stranger, waiting still.

"'I fought with them. I held them in my arms while their life's blood ran away from them on the hillsides. 'Do not weep,' they would tell me when they died. 'We are not sad, for we have given our lives for the people.'

"You see, that is how I learned. That is how I was trained, with the best of them, dying all around me. That is how I came to be what I am." He said it proudly, convinced in his heart of what he had become during the narrative. "This is the story I tell the boys. Whenever they falter, I remind them. If I could fight at thirteen with an exceedingly small frame, how can they be frightened now? No. They cannot waver. They must have strength in their courage if they are to fight with Memo!"

"To be sure," the stranger repeated gently. And Memo, having lost his inspiration, stared mutely in reply.

For a brief moment the moon came out from behind the cloud and turned them all gray and silver. They sat in silence, like pewter statues of themselves. Nothing moved.

"To be sure," whispered the stranger. "It is because I know of this that I have come so far to find you, Memo." He studied the drunken young man's face and decided to go on now rather than waiting till morning. "I have come to you, Memo, for nothing less than the bravest of them all will serve. No one but you, Memo, can handle the message I have come to deliver."

Somewhere in a remote corner of his mind, Memo heard the man's voice speaking to him and he knew it had come time for repayment of the debt incurred when he accepted the stranger's flattery. But it was too late to care, and Memo was too tired from walking fifty kilometers without stopping in his exceedingly small body.

"May I speak with you in private?" asked the stranger who called himself Beto.

"With pleasure," replied Memo. "You honor me."

"Bring us another calabaza of chicha," said the stranger. Memo lay still on his side, so the stranger made his way around the campfire and helped him to rise. Putting his arm around the younger man's waist, as if in friendship, he half-carried him toward the cave. To reinforce this charade, he smiled at the group. "We are good comrades," he remarked to them. Obediently, they nodded in answer.

With his right arm around Memo, Beto was now clearly seen to be smaller than Memo, but he was also clearly stronger. His left hand carrying a calabaza of chicha, the stranger departed into the darkness of the cave. Outside the boys covered themselves, one by one, and rested. Soon they were sleeping. After a while, the fire slept with them.

When the coals had grown gray and dusty, the stranger walked out of the cave alone, and sat quietly in front of the ashes. He sat looking into the forest, where he could see nothing. His eyes pulled down close and seemed to focus as if he were watching intently. There was nothing but darkness to see. Still, he watched the night with hard-burning concentration.

)

*Six* 🌙

By the time Memo woke the next morning and crawled hazily out of his cave, the sun had been long in the sky. An alarming rustle of activity reached his ears from the trees around the clearing—sounds of crashing and whistling and shouting, sounds like fiesta time. Only, it seemed to Memo, this was much louder.

"¡Madre de Dios!" he said, wincing involuntarily.

Memo lurched forward toward the campfire, looking for chicha, but there was none. In despair, he sank to his haunches and poured out the dregs of the morning's weak coffee; then, sipping slowly, he began grumbling beneath his breath. "Ayeiii, qué terrible. No tenémos razón." He shook his head and peered glumly toward the trees and the sounds of working laughter.

"What futility," he said to himself. "Oh, Memo, someday you will learn about drinking chicha with strangers."

But then the stranger was beside him, crouching just above him and peering into Memo's bloodshot eyes with a look of mingled intensity, questioning and respect that left the younger man confused. Memo found himself unable to meet the stranger's gaze.

"It goes well," commented the stranger. "Your boys are just as I thought they would be, good workers, eager to help the revolution."

Memo shook his head and dipped his face to the blue-speckled tin mug.

"You will be proud of them," continued the stranger. "When you are ready I will show you what they have done already this morning. They have accomplished much for so few."

But Memo only grunted and refused to look up.

"You will be remembered," the stranger said softly, speaking into Memo's ear. "When the guerrillas arrive they will know who has done this, and they will be appreciative. You will be respected, Memo. You will be one of them." He could have said "one of us," but he was not ready yet to drop the illusion that they were equals and confidants.

Finally Memo spoke.

"It is not reasonable," he said, still looking forlornly into his cup. "It is foolishness, man. You must know that. Why are you taking such pains to trick us like this? Why do you tease the boys into building huts when there is nothing to put in them?" But even as he protested, he dimly remembered the previous night's conversation.

"There will be, Memo, soon," whispered the stranger, still watching his prey as intently as a mountain cat, and with the same eyes, the same fixed purpose, the same concentration of a hunter poised for the spring.

"I do not believe you, man. How can I believe you?" Memo's voice was pitched high with tension.

"You believed me last night."

"That was last night."

"But today is no different."

"Of course it is different. Do you take me for a fool? Last night we drank and told stories and the whole world was open to us. We all believed it. Everything belonged to us then. But that was last night. That was liquor and firelight and good stories and good companions. Now, it is morning, and in the morning the mouth is sour and the belly is empty and I am wondering what we will eat next week, and how many boys will stay through the winter." Memo paused and took a swallow of the almost cold coffee. "Maybe tonight, maybe later, when we have found some food, when we have made a raid or held a discussion among us, then we will feel better and the war will be ours again, but, no, man, not in the morning."

"That is how it was, Memo," coaxed the stranger. "I understand that. I can see how you feel. But everything has changed now. Everything is different. Now the war has come. The men are marching to you, Memo, as we talk, and now we have no more time. We must be ready, Memo, for a man with no food cannot fight for the revolution."

"It is hopeless," protested Memo. "No one in the village will give you a thing—no food, no money, no supplies." Then he hesitated for a long time, gazing at the man called Beto with a strange, almost shy expression. Finally, he began to speak again, his voice soft.

"I have to tell you honestly, camarada, we are not much respected in the village. For their money we would be better in the terraces hoeing the maize crop or carrying pigs down the hill on our backs. I am afraid they do not much believe in the revolution, that it is real. You have to understand these people, man. For so many years there has been talk of the revolution, but nothing has ever come of it. Nothing ever changes from year to year, so it is only good for conversations. It is good for arguments in the hot afternoons, man, but it is not real. It is not real to them, and I confess, though I am a true believer, it is not real to me either."

"But, Memo, you have been there. You have fought with the best of them, remember?" goaded the stranger.

Memo ignored the taunt.

"I do not hold it against you," said Memo, "that you cannot tell the difference between what is real in the evening and what is real in the morning, because you are a foreigner and not a native of these parts. But if you had lived here as long as I have, you would know that it is a very poor village. There is hardly enough to sustain the few that live here. I tell you, hombre, Pasaquina is of no importance to the world. I know that for a fact, man, and I have lived here a long time. No man in his right senses would bring the war to Pasaquina."

"Perhaps you are right, Memo," agreed the stranger, never losing his calm, quiet intensity. "But the war is coming all the same. I have just left them and they are marching this way. The guerrillas must have a place from which to fight. They must have a safe resting place and food. It must be hidden and near the top of the mountain so that when the soldiers come up the hill the guerrillas can pick them off from hiding places. You see that, don't you, Memo? We talked about it last night. Last night you agreed that it was wise."

But Memo only grunted.

"This morning you will take me into the village,"

the stranger continued with friendly firmness. "We will determine what is available there. We must know what there is that we can use. When you have finished showing me the village, we will reconnoiter. You will show me the crops, the streams, the lambs, the cattle, the pigeon coops, the herbs, the motates, even the lair of the conejos. You will show me everything, Memo. It will be a long day, for I must know it all."

Memo shook his head and appealed to the stranger. "They will never give you anything, man. I've told you that."

The stranger ignored his protest. "And tomorrow, Memo, we will do as I have promised you. I will take you to see for yourself how close they are, the boys of the revolution. On the other side of the mountain, you will be able to see the enemy's advance. Then you will know for yourself why we must hurry."

Memo groaned. "It is a long walk, verdad? It is very far away?"

Slowly, the stranger shook his head. "No, chero, it is not far away."

"But we can hear no guns. If it is so close as you tell me, why can't we hear the guns?"

"Because, Memo, as I told you last night, for the time being they are well hidden. They are safe. The army does not know they are there. But every day the army draws closer, and they are many. It is only a matter of time before the retreat must begin. The men are tired and sick, and they have not eaten well all summer. It is better if they withdraw in silence to a place they can defend without difficulty. If this is possible, then they can rest until spring. They can heal to fight again. You understand this, Memo?"

"Sí, but this is not the place, hombre. I tell you it is a very poor village, and not generous to strangers."

Beto nodded, and Memo felt he must have persuaded him, but the stranger only replied, "Can you be ready soon, Memo? We must begin."

Seven

Father Herrera had not yet decided what to do with Sister Magdalena, whose presence remained an embarrassing novelty to him. The problem had distressed his digestion at supper and at breakfast, and hence all he could think about was righting matters before lunch time, which meant that she must be sent out somewhere to see somebody for something so that he could eat his lunch in privacy.

Since Rita's attention was mostly on their new resident anyway, the priest assigned her to take Sister Magdalena into the village and introduce her to the people. Why this should be done he was not sure, but he vaguely remembered that nuns visited people. Since there was no orphanage, no hospital, no clínica, and no colony of pagan Indians, he felt at a loss about the matter and returned instinctively to his old habit of trusting God and a woman's intutition.

"The Lord will guide your steps," he told Rita as soberly as possible and then shuffled off in search of an excuse to remain uninvolved, huffing and wheezing officiously. He retired to his room, opened his book at random, and peered through the shutters. When at last he saw the two women pass through the courtyard gates, he breathed a sigh of relief. "I must write a letter to the bishop," he wheezed. "Immediately." His stomach growled ominously, so he put down the prayer book and stretched out on his bed. "I will write him immediately tomorrow," he sighed. "Tomorrow when I have recovered . . ." Within three minutes Father Herrera was snoring blissfully, his rosy cheeks as peaceful as a baby's.

Rita, however, had no intention of visiting the parishioners. Her course had been set from the moment Father Herrera made the suggestion. Rita was no peóne, and she was not about to share her treasure so easily; if the villagers wished to speak to the sister, they could come to her, and she, Rita, would make inquiries on their behalf. She was temporarily in possession of the most sought-after object in the village, and she was not about to relinquish her prize for no better reason than an absent-minded order from Father Herrera. Rita felt in her heart that a little showmanship at a time like this would go a long way toward overcoming the whispers and rumors that had grown so popular with the villagers since the birth of her second son. Just let them say Rita was unclean, when they saw her with the treasured Virgin.

The corners of Rita's mouth turned up happily as she and the sister stepped outside the gates into the warm morning sunshine of the village square. Rita carried a woven string matata in the crook of her right elbow and laced her left arm through the arm of her companion. Though the two women spoke very softly and observed all possible decorum, anyone could see that they exhibited a strong sisterly affection for each other.

Dropping their tattered rags, two boys who had been polishing tables at an outdoor café hustled quickly into the shadowy interior in search of the proprietor, their eyes black and blinking like squirrels'. "Come and see," they hissed. "Look for yourself."

The stout women sweeping in front of their stalls full of round cuajada balls, milk, and oranges, stopped like the boys, put down their brooms and hurried off in search of their mothers and children and grandfathers—everyone, in fact, who had not already gone up the slopes to tend the crops.

"Come and see!" they whispered behind their hands, sweat like tears on their foreheads in the morning heat. "You won't believe it! You'll see who she has chosen for her companion. I tell you, it is a sign from God."

So they looked. In tiny groups beneath the worn canvas awnings of the cafés, from shaded adobe doorways, from corners of sun-bleached alleyways, from beneath the ceiba trees, from windows with yellowed lace curtains, from all around, they looked. Silly girls in scarlet headscarves with fat babies in their arms giggled with embarassment. Children in ragged shorts and hand-me-down shirts with no sleeves ran barefoot from the narrow streets, laughing and calling, to squat like monkeys on the dry fountain and watch.

While Rita and Sister Magdalena were passing by, the onlookers did not speak, but only nodded courteously and dropped their eyes. Once the two young women had passed, however, a storm of whispering broke behind them.

Rita managed not to smile or even speak to her

companion about what was happening, though she took it all in. She kept her voice light and continued her proud narrative, which was a sort of travelogue, but containing much more interesting information, such as who had died in this house, and how many babies this señora had borne, how many were left and how big they had grown, and which shops cheated even the priest. Not to mention—here Rita's monologue was slightly shy—the woman who owned the small cantina off the square and the terrible diseases that mysteriously attacked girls who came to work for her. Patiently, Rita toured Sister Magdalena through the entire village. They strolled arm in arm up one narrow, twisting cobblestone calle and down another until, by the time the sun was high, they were moist with sweat, pleasantly exhausted, and back at the square once more.

"Would you like un sorbete?" Rita asked.

"Oh, yes!" Sister Magdalena, unable to contain her excitement, almost clapped her hands. Then she remembered. "But I have no money."

"Never mind," said Rita, smiling proudly. "I have a little." She displayed a small pouch of stained red leather, laced tight at the top by a thong. "My food purse," she explained. "For shopping."

"Oh, but then we mustn't," protested the girl, the teachings of the convent still fresh in her mind.

"Of course, we will," said Rita, laughing gaily in her heart but outwardly maintaining her calm superiority, for the two women were the center of the entire village's attention. "There's always plenty. There's more than enough. I often treat myself to an ice cream when I do the shopping."

Sister Magdalena smiled at her, wishing for an ice cream but not daring to encourage her.

"Come. We'll just have a little rest from our walking before we market."

So they bought cinnamon sorbetes and sat side by side on a bench in the square to eat them. It was the greatest triumph in Rita's life. It was more wonderful an honor even than when she had been invited into Father Herrera's home and bed. She felt herself swell with pride and joy, and glanced over at her friend. The girl was concentrating on the ice cream with a look of such awe and pleasure that Rita was reminded of the pobres of nine or ten who saved or stole from their fathers a peseta and tasted ice cream for the first time in their lives.

Sister Magdalena's round, dark eyes were stretched wide and seemed to glow with an inner luminosity that could only have come, it seemed to Rita, from a pure and ecstatic heart. And suddenly, watching the girl's serious face, which shone as if from a heavenly vision, Rita knew what it was to be thoroughly in love.

For the first time in her life, in a yearning of love and pride she had not felt before, Rita's heart went out completely from herself. And when her heart was gone altogether from her breast and resided in the smooth countenance of her beloved, she found that it would not come back again. It was the most wonderful feeling she had ever experienced. Yet no one watching the two oddly matched women as they licked their cinnamon sorbetes in the sunshine would ever have guessed the transformation which took place in Rita. There was nothing to be seen on her face but a brief flicker of wonder.

When they had finished their cones, they started off again, but this time Rita was in the lead, with Sister Magdalena following slightly behind. Rita was firm and businesslike now, her nose slightly elevated, her eyes hooded, her hands folded firmly across her stomach. The string bag dangled to her knees. They came first to the produce vendor, but Rita only paused, shaking her head in disgust at the woman's offerings, then proceeded to the next stall. When they had passed all of the stalls, and Rita had shown her disdain for all of them in turn, they ambled back to the center and paused in front of the woman selling harina de maíz in ten-pound manta sacks. Rita stood for a moment, looking at the sacks with utter contempt; she prodded them with her fingers, mashing, turning, then pushing them away to show their inferiority. At last she fixed her eyes on the vendor, and the bargaining began.

"¡Cuánto es!" she demanded. It was not a question.

Buying the flour for pan francés required five minutes. Next came a performance which Sister Magdalena and the villagers could all appreciate, for it demonstrated beyond question the wealth, and consequent superiority, of the church. Rita bought dried figs, green bananas, dried salt fish, pepescas, one plucked chicken, a slab of pork, four pan dulce, and even a small chunk of leche de burra. It was true, of course, that Rita could never have purchased so freely from her grocery purse had she and the vendors not been showing off for the sister simultaneously, but it was also true that everyone in the whole market, including Sister Magdalena, Rita, the old women, the lame men, and the grandfathers, was impressed and pleased beyond measure and felt a warm, solid satisfaction in their stomachs when the purchasing was completed. They all smiled freely when Rita's string bag was filled, and nodded after the two young women as they walked, arm in arm, back toward the square.

)

*Eight*  🌙

Memo had not felt well all morning. Whether it was from the chicha he had poured on top of the cantina's chaparro, or from the temperature, which seemed oppressive, or from the embarrassment for his village that he felt in front of the stranger, he could not say. But he knew he did not like the villagers seeing him with the stranger almost as much as he did not like the stranger seeing the village. No good could come of it. Memo did not wish to be at the center of trouble, and he could feel trouble coming, like a man can feel drunkenness approaching when he has passed his limit.

The trouble coming felt like drunkenness, in fact. It felt like vomit rising up in the pit of his stomach. Memo found himself sweating beneath his straw hat, which he had worn in place of his moth-bitten beret, hoping to avoid recognition. That was a foolish idea, of course, for the stranger with him was like a flaming torch. But he felt, nevertheless, that the gesture might somehow reach out to his fellow villagers and proclaim that he was one of them.

Memo had always dreamed that when the fighters came to Pasaquina he would be proud. He had dreamed of the day he would strut manfully through the square in the company of a tough, weary guerrilla and show them all. But the day had come and now he only wanted to pull his hat lower over his face and hide himself from them. He refused to wonder why this was so, why he felt shame in the stranger's presence rather than pride. He only walked with the stranger and spoke to him in a brusque voice, detailing the businesses, the houses, the stalls, just as he had recited the multiplication tables for the old padre years ago when there had been a school.

If the stranger noticed Memo's discomfort at all, he gave no sign. He moved through the streets of Pasaquina in his heavy black combat boots as smoothly as if he were gliding on water. His face was calm and his voice was soft. When Memo protested, or suggested they return to the forest, he only pretended he hadn't heard and asked a new question. So Memo, as the morning grew old, was forced to walk one more block, turn one more corner, recite one more lesson, be seen one more time at the stranger's side. His head hung lower and lower on his chest, and his voice grew more and more hoarse, as if he had caught a fever.

But still, the stranger seemed not to notice his companion's misery and moved as deliberately as ever past each house, each shop, each barroom, each café in the village. He never asked to be introduced to the villagers, though they gawked at him wherever the two went. Once a group of children even sang derisive songs behind their backs until Memo was forced to throw stones at them. But the stranger only stuck to his course, pausing to listen as they passed each house and the details of the household were given to him. When he felt the information was too scanty, he would prod Memo until he was satisfied, then nod silently and move on. Increasingly, Memo felt there was something nasty, perhaps even sinful, about the whole thing. Certainly no good would come of it.

Finally, toward noon, the stranger turned back toward the center of town. He had exhausted every nook of the village, so Memo was hoping that this was the end of their distasteful excursion. But when they reached the square, the stranger said, "We will pause now to eat." He pulled some colones from a pouch he kept inside his shirt and sent Memo for tortillas and beans. Seeing the amount of money he kept on him, Memo suggested a beer at a café, but was sent on to the vendor with a wave of the stranger's hand.

While he waited, the stranger sat himself against an empty stretch of wall and watched. The villagers were also watching him, but they were furtive. They looked quickly and then looked away again, or peered past doorways, or squinted up from under their hats. The stranger sat and watched openly, though what he saw around the lazy square could not possibly have been of interest to anybody, as the villagers knew, save themselves. They wondered what had brought him to Pasaquina, who he was and where he had come from, but these questions were of only a little interest. The more burning question in their minds concerned the curious—and certainly significant—fact that this stranger was following almost the very same path that, only a little before, had been trod by the magically beautiful young nun, whom most of them called in their minds, automatically now, the Virgin.

Memo returned with their food, and the two men ate silently. When they had finished their repast, Beto, the stranger, inclined his head in the direction of the church and said, "There is one last thing I would like to see. There."

"¡Santa María!" Memo hissed. "You! In a church! Surely you don't want to make your confession."

Then the stranger laughed out loud, his head thrown back, his lips splitting apart his brown face, his strong clean teeth shining. He slapped Memo on the back, which made Memo jump as if the blow were intended seriously.

"Memo," he laughed, "how you entertain me! What a good joke—confession!" and he continued to laugh loudly for what seemed to Memo a very long time. "No," he said at last, "I don't need a priest, Memo, but a garrison, a fortress. You never can tell when such a place might be useful."

"The church?" Memo was dumbfounded.

"Some of these old church buildings make fine garrisons. Lots of rooms, built like a mountain, I tell you. Some I've seen have walls three feet thick. They couldn't withstand a missile or a good tank battery, but they'll keep out the rain!" He laughed, immensely pleased with himself. "Sure, Memo, it might be worth something to us. I want to see the church, the padre's house, everything."

"Sí, hombre," Memo sighed and shuffled across the square, hoping this foolish thing would be the last. The stranger followed, his pace impassive, his eyes alert.

The courtyard gates were open and the square, dusty space stood empty, except for a cluster of multicolored hens scratching the dirt in a far corner. From the back of the courtyard came the sounds of women laughing and chattering, the clank of pots and pans. Memo and the stranger stood just inside the walls for several minutes while the older man surveyed the buildings intently.

"That door—where does it lead?" he asked Memo, gesturing toward a door that opened off the covered walkway and into the back of the church.

"How should I know?" Memo answered impatiently. Although he did not like to admit it to himself, the stranger's investigation of the church was making him feel even more uncomfortable than their tour of the village.

"Then let us see," said the stranger calmly, and he started across the courtyard, with Memo following reluctantly a pace or two behind.

As they passed within sight of the kitchen, Rita became aware of them and stuck her head out through the door, wiping her soapy hands on her apron. Behind her, in the shadows, the older woman Father Herrera had brought in to help pulled herself back after one quick peek and kept out of sight.

"What do you want?" snapped Rita. "What are you doing here! The padre is taking his siesta. He can't see you now."

"No, hija," said Memo, "don't trouble him. I only wanted to show my friend the church."

Rita glanced at the stranger, noticing immediately his fine boots and soldier's clothing. "Who is he?" she demanded.

"A friend," said Memo.

"What kind of friend?" she insisted, looking fiercely at the stranger.

Memo shrugged helplessly and looked over his shoulder at Beto, who was standing quietly behind him with a faint smile on the corners of his lips for Rita. "A soldier . . . a revolucionario."

"What does he want here?" Her short, round frame

was as straight and stiff as a wooden plank. Her nostrils flared, and her deep-set black eyes menaced the man.

"I told you, hija, just to see the church."

"You wish to pray?" She addressed the stranger, her voice skeptical.

"Yes," he replied, the small, enigmatic smile still at the edges of his lips.

"Take off your hat," she said sternly.

"I will," he answered evenly, and Rita, still somewhat grudging, began to lead them across the courtyard toward the back entrance to the sanctuary.

"I wish to pray in private," said the stranger softly to her back.

Alarmed, she turned to him. For a moment there was a look of fear on her face.

"Is the church not open to everyone?" he asked in a faintly mocking tone.

Rita glanced toward the sanctuary door, her brows raised in worry, then back toward the rooms opening onto the other side of the courtyard. She saw that Sister Magdalena's door was closed and relaxed slightly, thinking the girl was napping.

"All right," she agreed. "But take off your hat and be silent. You must be silent. It is a house of worship."

"I will," he said again. "Thank you."

She watched the two men enter the church, waiting to be sure they removed their hats, and then watched as the big oaken door closed behind them. Rita understood in her depths that her home and her church were threatened by these men, that they were interlopers and should not rightfully be allowed into the sanctuary, but she could not understand why. Father Herrera listened to the confessions of the guerrillas and gave them blessings regularly. He explained to her that this was his duty, even though he thought himself that their ideas were foolish and dangerous. Rita remembered the priest's remarks to Dueña Isabel about the wickedness of revolution.

But, she reasoned, the muchachos of Pasaquina were not truly guerrillas, they did not actually fight, and so perhaps Father Herrera was tolerant of them. This stranger, however, was another matter. He was not a muchacho, that was plain. Rita stood there in the sunny courtyard, pondering the fear and foreboding she felt. Then, catching sight of the children, she yelled threateningly at them before stomping back to the pots and pans waiting in her kitchen.

In the sanctuary, only a few candles burned on the side altar, doing little to illuminate the dark room. The big double doors at the front were pulled shut against the afternoon heat, and the windows, which were very small and high up, admitted only slivers of light that did not even reach the thick duskiness below. Because of the darkness, and because she was dressed in black, and kneeling, which made her very small indeed, Memo and Beto did not at first see Sister Magdalena. When, after a moment, Memo caught sight of her, he stopped short and gasped. Though he could not see her face, he knew from her diminutive size that this must be the Carmelita the village had been talking about. Even in the dim light he could distinguish the delicacy of her small, white-skinned hands, and this by itself was enough to make him nervous. He felt the old pull of superstition and had to hold his fist at his side to keep from making the cross over his breast.

Stepping up to Memo's side, Beto saw her too. He stood for a moment, letting his eyes adjust to the gloom, and then he looked carefully at her face, her hands, her small, delicate body. She seemed raptly attentive to her own thoughts, and unaware of the men staring at her. Beto's eyes grew dark as he looked at her, and his face suddenly relaxed into softness. At first, he simply luxuriated in the unexpected sight of such beauty. Then a sadness crossed his features, a look of pain so alien to his face that had Memo seen it, he would have thought himself looking at another man. But, quickly, it was gone.

Beto took Memo's sleeve and, pulling him behind, stepped quietly into a side aisle. Carefully, he examined the structure of the church, walking slowly up the aisle, his fingers touching the stucco walls, exploring. Near the back, on the left side, he discovered an area of newly patched plaster. He took a small folding knife from his pocket and gently picked away the scabs of plaster; he nodded, satisfied, and turned across the back of the church. Coming after him, Memo saw that the stranger had uncovered a small area of the rock slab construction hiding beneath the plaster. Memo also nodded with satisfaction.

They were silently coming down the right aisle, past the confessional, when she became aware of them. Her prayers finished, she had gone to the candle rail to strike a candle for her family—for those dead and for Roberto, whether he lived or not. When she raised her eyes, she saw movement coming toward her. Her fingers flew open, dropping the match they had been holding to the candle wick, and she cried out like a frightened hare.

The stranger held out his hand as if to stop her, and said, "We have only come to pray."

"Forgive me," she said, "I did not hear you. You walk so softly."

"It is best to walk softly in God's house." His voice was quiet, as still as her face, as smooth and unrippled.

As they drew closer she noticed the costume he wore. "A revolucionario," she said to herself and a flicker of light came into her eyes. "He is one of the people," she thought, and then she looked past the stranger to Memo. He too was dressed vaguely like a guerrilla, but looking at his face did not spark the small happiness, so she looked back at the stranger.

Beto paused. He did not want her to interfere with the remainder of his explorations, so he spoke to her with that cunning voice he had developed to gain his wishes from others. "I hope we have not disturbed your prayers, sister."

"Oh, no." She flushed, suddenly naked, or feeling vulnerable in that same way. "I have just finished."

"Your candle," he said, gesturing toward the small, still-unlit cylinder of wax. "May I give you a match?"

She looked at him directly then. One quick look into his black eyes. But she could not look long. "Oh no, I have one, thank you." She took another from the pocket of her habit, lit the candle, made a small curtsy, and backed away from them. "Excuse me," she said, "please excuse me now." Quickly, she was gone.

Memo stared after her, by now having forgotten entirely the purpose of their visit, for he had had a glimpse of her full face. "Yes," he sighed, "she is indeed the same. She is just as they say."

"What?"

"Didn't you see her? Didn't you see her face?"

"What?"

"She is the image of the Virgen de Guadalupe. Didn't you see?"

The stranger paused a moment before answering. When he did, his voice was uncharacteristically gruff. "I don't know much about virgins."

"It's a miracle," whispered Memo. "It truly is."

"Come on, man," said the stranger, pushing him into the small room off the altar. "There are no miracles. There is only what you take with your hands in this life. There is nothing else. Shouldn't you know that by now, Memo, a pobre like you who's done nothing but scratch in the dung for crumbs all of your life! Where's your mind, hombre, that you let a smooth face set you blathering like an old maid?"

With that he turned away from Memo and looked carefully around the small, windowless room, recording it in his mind with the rest of the information he had garnered that morning. Here was the other side of the door he had asked about in the courtyard. A locked chest and a small table were the only furnishings of the room, which was, in fact, the sacristy, where Father Herrera put on his vestments before Mass and where the chalice and other items for the ritual were stored. But to Beto's eyes the room revealed its usefulness for another purpose altogether.

When he had finished his inspection, Beto trudged back up the aisle of the church toward the double doors at the street.

"Hurry up!" he said to Memo. "We have all the fields in this area to cover by sundown."

"It is not possible," protested Memo.

The stranger didn't answer him but pushed out into the sunny square, leaving the great doors open wide behind him. Memo trudged along like a dog on a leash and shook his head in sorrow. "No good can come from it. No good at all."


Sister Magdalena found Rita sorting frijoles in the shade of the veranda. "Who were those men?" she asked.

"Nobody," said Rita, with a shrug.

"They were fighters, isn't that true? Fighters for the people."

"That's what they say," Rita sniffed, "but all they really are is rag-tag and bobtail who live in the mountains and play little boys' games because they are too lazy to work. That's all they are." She had been thinking about this point, and had concluded that it explained a great deal.

"No. Not the smaller one," said Sister Magdalena. "He is a real fighter."

"How do you know, child?"

"It's because . . . I don't know how, perhaps it's because he's so stern . . ."

"They think only of themselves," said Rita, whose position was growing more firm by the minute.

"No, not this one. I could see it in his eyes," said the girl wistfully.

"Child, forget about it. Guerrilleros make bad husbands."

"Why, Rita?"

"They don't like houses and they don't live long, that's why!"

The girl laughed. "Oh, Rita, what would you know about freedom fighters?"

But Rita wouldn't answer. "Forget it, I tell you.

Don't look at them. They are no good, son buenos para nada."

Sister Magdalena took a handful of beans from the basket and sat down to help. "You'll see, Rita, someday when the war is through, you'll see. They're fine men. They're noble men."

Rita laughed. "And you are an expert on men?"

The girl blushed. "You laugh at me, Rita, but some day you will see. They will make us free."

"There is no such thing as long as we are in these bodies," pronounced Rita firmly. "We are not free no matter who lives in the capital, because we must always eat, and to eat, you must slave. Maybe in heaven it is different, but here on this earth, it is useless to worry about freedom. It doesn't exist."

Sister Magdalena laughed.

But Rita kept on. "It may be that you think of your father's ideas and believe that you should feel as he did. But you are not a man. Men think they must do something grand and important. And as often as not the result is like what happened to your father." Rita's tone softened and grew sweet. "I do not want something bad to happen to my Luna."

They worked at their task in silence for a while. A black fly buzzed about Rita's oily hair. The girl watched her own pale hands, mechanically sorting through the beans. A red rooster tentatively pecked at the discarded beans. Out in the square, a goat boy and his herd passed through. The metallic clanking of their bells mingled with the sound of horses' hooves on the hard ground, the squeaking of wooden carts, the vendors' cries, and the squalls of waking babies. Then the beans were finished and Rita rose to return to the kitchen.

"But Rita . . ." Sister Magdalena spoke timidly.

"What?"

"Oh, it is nothing."

"No, tell me. Tell me your nothing."

"It's just that I can't find them bad . . ."

"They are not part of us. They are not a part of this village. They have chosen that, so that is what they have."

"But, it's, well . . . something about him . . . something about his look . . ."

"The older one?" The one Rita feared.

"Yes."

"I don't like him."

"Rita, there was something about the way he was, just for a moment . . . maybe it was that look in his eyes—the look of a man who knows he is going to die. I don't know . . . but there was something. I saw it! Something that for a tiny second reminded me so much of my father . . ."

"You are too romantic. They are only hungry. That is what you saw in his eyes—hunger. Just hunger and greed."

"Yes . . . I know they are hungry . . ."

Rita slammed the pan of beans on the wooden bench. "Oh, child, you are more trouble than all my others together! All right, if you will be a donkey, come on."

"What are you doing?" asked the girl.

"You are not convinced."

"Yes, you're right."

"Then follow me." Rita stalked off to the kitchen and slammed a kettle on the hearth.

"What are you doing?" the girl repeated, trailing after her.

"Sit down."

Sister Magdalena obeyed.

"Now, I will tell you what we are doing. I am making café, and then . . ." Rita's voice was silent. After a moment, during which Sister Magdalena noticed Rita blinking very hard, she continued. "Then, my little Luna, I will tell you a true story, and you will see that the guerrillas bring only trouble for women."

)

*Nine* )

It was not even first light the following morning when Memo felt the hard toe of a boot digging into his ribs.

"Wake up, man," the stranger was urging him. "Wake up!"

Memo groaned and rolled over, opening his eyes cautiously. The air was damp and chill and still dark as midnight; even the birds had not stirred yet.

"Hurry up. It's time to move out," Beto was saying. Memo looked up, squinting into the darkness, and saw that the stranger was bending over him, holding out a cup.

"Here's some coffee, friend. Sit up and drink it. I've put a little guaro in it to warm your stomach." The stranger handed Memo the tin mug and crouched in front of him on his heels. "Now get up, we have to move."

"¡Ayeeiii, cabrón!" said Memo. The scalding liquid seemed to lift the skin from the roof of his mouth. "It's hot! Why didn't you tell me?"

"How else do you drink coffee?"

"Up here it is never hot."

"That is because you sleep so long in the mornings," the stranger laughed. "Hurry, man, get it down. I have food in my pack. You can eat while we walk."

"What nonsense is this? You said it wasn't far."

"It isn't far."

"Then why must we leave so soon? It isn't light. We'll break our legs on this mountain with no light," Memo protested.

"I know the way," answered the stranger calmly. Even though Memo could not see the man's eyes, he knew they were looking at him, straight and steady, waiting.

"I hope so," Memo answered grudgingly. Then, still grumbling, he rose and followed the stranger out of the cave. At first he could see nothing and blinked helplessly in the darkness, but after a moment his eyes picked out the huddled forms of the sleeping men, and the small glow of embers that the stranger had roused from the remains of the campfire. When he turned back to look, the stranger was already twenty feet ahead of him, at the edge of the clearing near the side of the mountain, and was about to enter the laurels and the low-limbed cedars. Hurriedly, Memo followed. He had no wish to be caught alone in the forest while it was still dark. Although he did not actually believe in the vampires that were said to lurk there at the top of the mountain and down the other side, nevertheless, one should not be foolhardy.

For the first hour before daylight, it seemed to Memo they did nothing but scramble over boulders. If there was a trail he could not distinguish it by its feel beneath his feet. Once, he slipped on an incline covered with loose pebbles and thought he was going into a crevasse, but he found himself sliding only ten feet or so and ending up with his nose in the dirt and his back against a cedar trunk. Another time when he slipped, which was all he seemed to do—slipping backward, he thought, more than he climbed forward—he reached out in a panic and grabbed a handful of thorns. The stranger ignored his troubles entirely, making no move to help him. And whenever he tried to gasp out a protest, the stranger silenced him harshly.

"Come on, hombre, we've no time to lose. It's growing light."

But Memo could not see the light and knew it wasn't there, at least on their side of the mountain. He knew that their side of the mountain was on the east and so suspected that either the stranger had eyes in the back of his head or was taking him off into the blackness to slaughter him. In his musings he forgot to listen to his feet and tripped against a root, falling forward and cracking his shin against an invisible rock ledge. The pain was so intense he thought he had broken his leg and cried out accordingly.

"Hush, fool!" the stranger reprimanded him. "Don't you know they never sleep?"

Memo was so bewildered he felt tears spring into his eyes.

"Hombre," he said, "why are you doing this to me? I have never offended you. Why are you torturing me like this?"

Though the sky was still dark, it was not as dark as it had been before, so that when the stranger turned to him, Memo could make out the shape of his face and a blur of black where his eyes were, but it was too dark to see the expression in them. Memo knew it didn't matter, though, for the man had the eyes of a snake and there was nothing to be gained from seeing them.

The stranger touched him on the shoulder. Memo

winced. "My friend," the stranger said, "at the base of the mountain there is a river. Between here and this river are many trees, very thick. In those trees are the guerrillas. They are without supplies. They have run out of food and medicine. They have very little ammunition." The stranger stopped speaking, and Memo thought that he was gazing very far away, as if he could see the guerrilla camp. Then the calm voice continued. "On the other side of the river are the enemy. The army certainly, and perhaps the Guardia. They know that the guerrillas have crossed the river, but cannot see them in the trees and do not know whether they have camped or moved to the north or to the south. The soldiers are either awaiting orders or pausing to plan a new strategy for attack. I do not know. But either way, when they begin to move again, these boys beneath us in the trees must be gone."

"Are we at the top, then?" asked Memo.

The stranger ignored him and continued, a low quiver beginning in the bottom of his voice. It was the only sign of passion he had yet exhibited.

"Memo, if we are to win this war, we must use trickery. The enemy does not need to hide. They are supplied down the open road on the main highways, and if their men are shot in the process, they are simply replaced with more. Their numbers continue almost endlessly it seems, for they are increased by conscripts from the cities and mercenaries from many places. But we are only a few, and our men are volunteers—like you, Memo. So we must at all costs preserve ourselves, you understand?"

Memo only stared at him.

"We have to take our supplies clandestinely; we

have no airstrips surrounded by mine fields and barbed wire fences; we have no headquarters with barracks and warehouses and permanent telephone lines, so we must make our headquarters where we can and cut airstrips out of the forests and valleys where they will not find them."

He waited for Memo to nod understanding, but Memo did not, for it was not so obvious to him. Memo could not envision all these things. He did not know what lay beyond Pasaquina. Nestled at the base of the mountain, with its terraced, hill-climbing fields and its brushy valleys and its meandering streams, Pasaquina seemed the world to Memo, though he knew in his mind that it was only a small part of something much more complicated. He could get to the road going north, but in their raids they had always gone the same way and only so far; he had no idea what lay at either end of the road or even on either side. It had never been important before, and didn't seem so now, despite the stranger's words.

The stranger seemed almost to read Memo's thoughts, and he went on as if to demolish them. "Once they reach this side of the mountain, the boys will have entered, so to speak, nothingness. Here there are no wide roads for bearing trucks and tanks; there is no industry, no commerce but village bartering, and hardly enough agriculture to pay the taxes and support the population, so there is a whole valley beneath Pasaquina that has been forgotten about. It is isolated from the world, cut off; it might as well not exist."

Suddenly, hearing the stranger's words, an idea came to Memo which was so startling, and yet so clear, that he felt himself shaken as though by a tremor of the earth.

*For us,* he thought, *Pasaquina is everything. For them it is nothing.* His earlier feeling of shame in the stranger's company, which had seemed vague and inexplicable then, made sense now. Instinctively, he had realized that the stranger looked on the town and the people as if they were toys, to be played with and broken.

Beto's voice continued, smooth and persuasive. "So you see, Memo, that is why Pasaquina will make a perfect base camp for us. The clearings below the village are practically a landing strip already. And helicopters can come in from the northeast, keeping low behind the ridge of the mountain, without ever being detected. We can receive supplies, store them in the trees on the top of the mountain, and so be ready for surprise attacks and ambushes on the enemy when they begin to move. From the top of the mountain, we have the tactical advantage, for they cannot move in any direction without being observed. Then, once we are resupplied, we can take them without difficulty. We have been promised mortars, Memo. Think of it. Do you know what mortars from this vantage point could mean?"

Memo shook his head.

"Even though we are only a part of the fight, this could be the turning point of the war, Memo." The stranger's voice was low and vibrant now. Memo's attention was fastened on him completely. "Pasaquina could very well be the cradle of our freedom, the birthplace of a real democracy of all the people."

"So?" whispered Memo, in surprise and confusion. He could feel something moving, clutching, inside his chest and stomach.

The stranger stared intensely at the half-lit sky. "It

is an awesome thought, to know we could be so near the end."

"It has been going on as long as I can remember," said Memo, stirred against his will by the promise of the stranger's words.

"It has been going on as long as any of us can remember clearly, Memo, but I think, I feel, maybe, we are very near the end." He paused to study Memo for a moment, then added softly, "And you will be there, part of it."

"Oh," said Memo softly, for he could think of nothing else to say. But gradually he was feeling stronger, and his mind was quickening. "And we are going to take this word to the guerrillas?" he asked.

"Yes. They will begin moving out today. They will follow us back up the mountain."

"Today?"

"During the night." The stranger's mouth widened, so Memo knew he was smiling. "And soon, if all goes well, Pasaquina will receive its first air drop."

"Food? Medicine?"

"No. Ammunition. Ammunition is more important to a soldier than food, Memo. More important even than medicine."

"What will we do with it?"

But the stranger only shook his head.

"The church," guessed Memo. "We are going to store it in the church."

The stranger chuckled softly in his throat. "Of course. And when the big guns come, they too will be hidden in the church."

"Ah," sighed Memo. He was no longer resistant.

With the mention of big guns, a thing he had never seen, and helicopters and airplanes, also things he had never seen, he was suddenly eager to follow the stranger. The fate of Pasaquina was forgotten. "Let's be going then," he said, and then added slyly, as if he had been the first to think of it, "If we wait until the sun is up there is always the chance they will catch the glint of light from our guns."

The stranger didn't laugh, but replied seriously. "You are quite correct, Memo. At this point we must use extreme caution."

So just before the first real moment of morning, they scrambled over the last ferocious boulders, crossed a ledge of rock and dropped onto a steep incline down which they could manage only a controlled slide. Memo wondered how the stranger had ever managed to climb it and thought there must be a longer, safer trail somewhere.

Soon, they had left behind the sparse, short cedars and entered a thick growth of laurels that were so similar to the ones on Memo's side of the mountain that he marveled to think two places in the world could be this much alike and yet not the same. Memo no longer feared the vampires. He no longer felt ashamed to be in the stranger's company, but began to look forward to their arrival in the guerrilla camp with the eagerness of a child attending a party of his elders. In fact, he could not remember now the way he had felt before, nor did he recall that, for a moment, he had understood the origins of his shame.

By the time the haze-covered sun had arrived they were well into the dense forest, so that only a dim light reached them; creepers and low, thick foliage filled the air and hid the ground beneath their feet. Although Memo could not see it, there must have been a trail of sorts, for only once did they have to hack through obstructing growth. As they walked, the sun began to penetrate beneath the trees, and the humid heat of the close morning replaced the difficulty of climbing in the darkness. Sweat poured down Memo's face and into his eyes. His breath came in gasps and his feet began to suffer, for his socks were wet with the sweat that seemed to pour inexhaustibly from every pore of his body. The pain in his shin returned and began to throb. Moreover, as if to test him thoroughly, mosquitoes smelled his sweating body and attacked his face and neck and bare forearms. Soon he wanted to claw at himself to scratch away the sting and the itch. He wanted to remove his gun belt and claw at his belly, and take off his shirt and scratch his back. Even his thighs stung and itched. He glanced ahead at the stranger and thought about asking to stop and rest, but the older man moved on through the brush as if he felt nothing at all, though Memo could see that his shirt was wet and his neck glistened. So Memo gasped and breathed as quickly and as often as he could and kept following through the trees.

By ten-thirty, they had arrived at a small plateau about a third of the way down the mountainside. Here the trees were thinner and the trail slightly more obvious.

The stranger turned and smiled at Memo. "We are almost there."

Memo nodded, trying to look composed.

After half a dozen steps more, they were abruptly halted by a quick movement just off the trail to their right. As if from nowhere, a figure appeared in their path, his face grim and his rifle leveled at their chests. But in less than a moment, the scowl was replaced by the happy grin of a very young man, and the rifle barrel arced away from them.

"¡Compadre!" the boy cried, stepping forward to embrace the stranger. The stranger embraced him with his free arm and boxed him lightly on his green cap. "Juancito," he teased, "you are up early."

With an almost adoring smile, the boy said, "We have been looking for you, Beto. Everything is ready."

"Ah, that is very good, muchacho," the stranger replied, laughing. Somehow, when Beto said "muchacho," it seemed an endearment, not an insult.

Juancito turned in the direction they were headed and whistled softly. He made the sound of a mockingbird. It was a complex and delicate warble, and Memo thought him very talented. Another mockingbird far behind them in the trees, fooled by the boy's imitation, answered. There was silence for a moment and then the boy repeated his call. This time it was answered immediately by the soft call of a mourning dove. Turning to the stranger and nodding happily, the boy was still. Once again the stranger patted him lightly, and then led Memo past him into the forest below.

The camp was not a camp at all. Memo saw that quickly. Though the trees on this level ground were thinner than on the hillside, it was still a forest, and no effort had been made to clear a space. At first, Memo had to look hard to see the men, for their dappled fatigues mingled easily with the bark and greenery of the forest. Also, they held themselves very still, though when he looked Memo saw them smiling. He had expected shouts and back-slapping and hearty salutes, but instead they moved slowly and quietly toward the stranger, as if the sound of a breaking twig could reverberate and be heard around the world.

The men, who were many different ages and sizes, gathered around the stranger and Memo. They all carried fine, well-oiled automatic rifles and wore good stout boots, but their faces were strained and tired. Some were bandaged about the head and hands, and one of them had his arm in a sling. Still, however, he carried his rifle, as did all the other men. Memo wondered at this; he had thought men only carried weapons when marching or fighting. And he wondered at something else as well. At the base of the trees, covered with branches, were stacks of equipment, well camouflaged. As Memo compared the size of the stacks to the number of men he could see, which he thought could not be more than thirty, he felt there were surely more supplies than they could carry over the mountain.

One of the tired, dirt-smeared men came forward out of the group and stood close to Beto. The two could have been brothers, so alike were they. It was not that their faces were shaped in the same way, but that they held the same story, which was grim and tragic and determined all at once. The new stranger had a long red welt that ran from his eyebrow to his jaw and puckered his cheek in the middle. It was a frightening scar, Memo thought.

"¿Qué tal, Beto?," the new one said.

"¿Cómo estás?" answered Beto.

The man smiled and shrugged and they laughed together as if he had made a joke. "¿Y tú?" he asked Beto in return.

"I," joked Beto, "I am as well-fed as a gringo, and as happy as a girl on her wedding day."

The other man laughed and nodded. "Then you have good news for us, Beto?"

"The best. It is everything we thought. It sits like a plum on a limb and waits for us."

"And the people of the village?" asked the new one.

Beto shook his head. "The only men in Pasaquina live on the side of the mountain with my good friend Memo." He patted Memo on the back to include him in the conference. "¿Es verdad, Memo?" he laughed.

Memo nodded stupidly, not knowing what they meant.

The new one grinned at him. "Welcome, Memo, welcome to the revolution."

"I have been a freedom fighter all my life," Memo answered self-consciously. But his words seemed hollow here in this strange camp on the other side of the mountain.

"Ah, like me," the new one answered. "They tell me there is another way to live, but if there is, I have never known it. He laughed softly. "I sucked blood from my mother's tits the day I was born. As you did, Memo."

Memo shuddered inwardly at the thought, but laughed as he was expected to.

"This man has been fighting so long, Memo," Beto told him with mock gravity, "that he has forgotten who he was. So far as anyone knows, he has spent his whole life in the hills shooting at the army. He thinks the human race propagates itself with guns. For every soldier he shoots, a little rebel is born under a bush in the forest, no?"

The new one laughed, his face reddening slightly.

"He has forgotten his name," continued Beto, "and since none of us has ever known it, we call him whatever we feel like. Sometimes we call him Miguel, sometimes José, and last week one of the boys said he called him

Lupita!" At this they all roared. "But he didn't mind, did you, amigo?"

The new one had turned his whole face red with laughing, and now the scar stood out like a white ribbon.

"You can call him anything you like, Memo. He has no name we can remember, so you can give him any name you wish. What would you like, Memo?"

Memo was confused and only shook his head with a sheepish smile.

"You have no ideas?" asked Beto. "Well, then, in that case, today we will call him Raul. That is a good name —Raul. Yes, I believe we will use that one today."

Raul stuck out his hand, and as Memo clasped it he noticed that the palm was hard with scars and the fingers stiff. "Come then," said Raul, turning back to Beto. "Rest awhile and tell me what you've found."

The two moved to a smooth area and squatted in the dirt, Memo and the others following. Beto wiped the dirt with his hand and then, taking a stick, began drawing in the smooth dirt. From behind him, Memo saw that he was making a map of Pasaquina. He marked the square, the streets that led off it, the alleyways, the winding trails, and the fields around. He showed the streams, the hills, the outcroppings of rock stretching further off than Memo had ever been. Once again something stirred in Memo's mind, seeing Pasaquina drawn there in the dirt like a game that children would play. Could these lines be his village, his mountain? Memo shook his head, as if a fly were buzzing near his face.

At last, Beto drew in the church and marked it off with an X. The men studied the map silently and then wandered off to prop themselves in the shade of the trees, their heads leaned back, their eyes half-closed. For a while Beto and Raul spoke together in low voices, but Memo swiftly became too sleepy to listen; their words ran together in his mind and seemed to float in and out of his hearing. Finally, Beto stood up and rubbed out the dirt map with his foot.

"Come, Memo," he said, "take some food and then we will rest. We do not leave until twilight."

Memo nodded. "The sun on the gun barrels."

"Lie here," Beto said, indicating a patch of fern at the base of a tempesque tree. "Sleep well. I will wake you when it is time."

Memo pulled off his pack and reclined on one elbow. As he ate a bit of cheese wrapped in a tortilla, he looked more carefully at the men. A few of them talked together in small groups, their voices low. But most sat alone, staring off into the distance or dozing lightly. He studied their faces as carefully as he could, to see if he could tell what distinguished them. Some were young, some were old, very old, like grandfathers, some were of indeterminate age. Some had the high cheekbones, hooked noses and dark skin of the mountain Indians; some had the broad, flat faces and wide noses of the plains; some had moustaches and beards and some were shaven clean. There were many physical differences. Yet somehow they were all the same, and Memo could not see what made it so. But he felt as if he were with men from another world who had only taken the bodies of real men to imitate them. That is how separate he felt from them. And as he chewed and swallowed and watched them, he thought he saw that they were, though the same, also separate from each other, as if each man had a world of his own which encapsulated him like an invisible membrane. Perhaps that was what made them the same—that they were separated.

Memo had never seen real fighters before and this was not what he had expected. He had thought they would be gruff and hearty, big and boisterous, peculiarly manly men. But as he watched them relaxing, he saw that they were just the opposite of what he had been looking for in his mind. They grew so still and so alone that he thought they had crossed over into the land of the dead. It gave him a cold feeling of fear to think of them as phantoms. How could this stillness win the war? There must be roaring and crashing to fight an army. But these men, he thought, were like the dead. Only the dead could stare unblinking when there was nothing there.

And he was right, though he could not understand his own realization. Something that they had been once before was dead. The names of the boys and men they had been before the fighting were no longer true, for the people they had been were no longer there, and even the memory was gone. What had been real in their homes and villages and shops and schools no longer existed, since even the memory was gone. In its place had come the stillness and the waiting. At first it had not been so, but then, as the counting had gone on, and the number of shattered skulls, and severed legs, and blood-soaked chests, and spilled intestines, and lost fingers in the forest, and the number of screams, and the groans and cries and calls for mother or sweetheart in the moment of dying, as the number of these things grew too long, they had stopped counting them and been still. But being still did not make the numbers diminish, so they had begun counting another thing and waiting. They began counting the number of hours until their own blood would be requested, until their own split hearts and mangled bodies would be called for. They began counting the number of hours and minutes and seconds until their own deaths, for counting life in terms of hours and minutes and seconds gave it a comforting quality of reality, while counting decapitated bodies and gallons of blood only gave life a surrealness which was impossible to bear. So they were separate from Memo and from each other because when death happened to their bodies, it would only be a reflection of what had already happened in their souls.

Ten

As nightfall neared, the guerrillas set quietly to work, with no orders given and little talking among themselves. Memo, who had been awakened not by Beto but by another man, the one with his arm in a sling, watched the men disassembling the camp. He did not know what he should do and so remained quiet, watching, until he was given his own pack to carry. Then, imitating the others, he shouldered the burden (wondering to himself if he would ever be able to get it over the mountain) and set off into the evening. Beto had insisted that they black their faces, which made Memo feel silly. The army, after all, was behind them and could only see their retreating backs, should they look. But Beto told him sternly that it was stupid and dangerous to take the enemy for granted, so Memo had blacked his face like the others.

The journey was slow and strenuous. The weight they carried up the steep hillside made the path treacherous, and they often slipped and stumbled. But never a word was said, and the pace set by the stranger was adhered to relentlessly. Memo would have cursed a million times had he been alone, but, wanting the good opinion of the others, he kept quiet.

When they neared the top of the mountain, Beto called a halt, gesturing the men toward him. They gathered close so that he could speak softly.

"This is the worst of it, amigos," he said. "We'll have to pull the load up on ropes; it is too steep and there are no footholds."

Memo remembered this stretch from the morning, when he had been so sure there was another trail. "Ah," he said to himself sadly, for now he realized there was not. This was the only way back.

"Couldn't we wait till morning?" he suggested, but the others ignored him. Winning the war was very hard work, it seemed to Memo; he had not thought it would be like this. Beto's voice, barely audible, continued.

"Tomás, Juan, take the rope and make it fast. There is a sharp outcropping just to your right when you reach the top. Attach the ropes there. Pablito, you will follow them up when they throw down the rope." Pablito looked disappointed but did not argue.

Tomás and Juan were both small and wiry, though Tomás was a youth and Juan a grandfather. They dropped their packs quietly in the dirt and slung their weapons over their shoulders; then Tomás took a coiled rope and fixed it to his free shoulder. Nodding silently to Juan, the younger man took the first handhold and soon was scaling the incline like a brown monkey. Below him the men watched in silence, straining their eyes in the faint moonlight. When he reached the top, Tomás slipped easily over a large boulder and peered back down at them. His blackened face might have been no more than another bump on the rock. Juan, the old one, was already on his way up. He spread his arms and legs wide and seemed to wriggle up the mountainside like a fly. Watching him, Memo marveled. In the daylight, and coming down rather than going up, he himself—a young man—had been terrified. Now this old man, this Juan, made it look like nothing. Memo wondered whether, in time, he himself would become so brave and skilled. Perhaps. *Or perhaps,* Memo thought with another part of his mind, *they are not brave but foolish.*

Juan was halfway up the slope when he slipped. There was no apparent reason for it, for he was moving slowly, clinging easily to the side as if he held no weight in his body, but something came between his foot and the earth and made him lose the grip of his black boot. So quickly that they missed seeing it, he came loose from the side of the mountain, though he clawed with his hands, his stomach, his mouth; he slid down on his belly for a moment and then lost even this adhesion, so that he fell to the side and began rolling. His cap came off and seemed almost to float down, landing near Memo's foot. Memo looked at it dumbly. Above him, Juan was still falling, as if in slow motion.

The side of Juan's head struck a tiny ledge and he grunted in pain. He flung out his arms to right himself, but it only made him flip first onto his back and then forward again onto his face. After that he came down quickly, like an avalanche, bringing with him rocks and dust and debris, until finally he stopped at their feet.

Beto had judged the precise spot where the man would land, and he was there even as Juan came to rest. He took Juan's head in his arms, and the man they called Raul grasped the flaccid legs. As if they had practiced many times, the two men moved Juan away from the rugged spot into which he had fallen and stretched him out on smoother ground. Another man was quickly with them, bending over. All was silent, except for the low groans from Juan's throat.

The new one, whom they called Julio, ripped off his shirt and began tearing it into strips.

"Water," he demanded. A boy came forward with a canteen. Julio swabbed at the gushing wound on Juan's temple and wiped the torn face as well as he could, then handed back the canteen. He looked up at Beto and spoke quietly while he began wrapping the man's head with the pieces of his shirt. "There's little I can do for him, Beto. It's bad."

"Will he last?" asked Beto.

The man shrugged.

"Does he have a chance?" insisted Beto.

The man considered a moment and then replied. "Here? Like this? Of course not. He will bleed to death before morning."

"But could you help him if we got him to the village?"

The man turned his face away, grim with resolution and fatigue and guilt. "It will slow us down too much. You know that, Beto."

"Will he regain consciousness?" Beto asked.

Julio nodded. "Probably."

"Then he will know himself to be dying, here alone, without his comrades."

"It is the oath we take, Beto. We all know when we begin what our fate may be."

"Yes, it is true, but this is an old man, a grand-father."

"And I too am a grandfather," Julio said roughly. "But I tell you, if it were me, I would not want my comrades to jeopardize themselves for my broken body. Nor would you want such a thing for yourself, I think, Beto. It is not like you to think in this soft way."

Beto said nothing for a moment. *My father would have been like Juan,* he thought to himself. *If he had lived.*

"It does not matter," Beto said then. "They've had enough for this season. They are not pursuing us now."

"If you do it for one, Beto, think . . . you must do it for all. What will happen in battle?"

Beto looked at the torn form lying on the bloodied dirt and then at the other men, waiting mute, not showing opinion on their faces, only respect and trust. He shook his head.

"This is not a battle. With luck, we can manage. We must try. We'll take him to the village."

"Pablito." He gestured to one of the boys. "Get up to the top."

The boy jumped forward, waved his arm to the black monkey face at the top of the mountain, and began climbing. Memo watched him with new admiration. To climb so surely after a man had just fallen. "What courage," he said under his breath. But the others were not watching. Beto had taken a long sharp folding knife from his pocket and was busy stripping the canvas pack which had been Juan's. Two other men had gathered a few slender saplings from nearby. They took the strips of canvas and wove them together with the strong flexible saplings to make a hammock. They attached three loops of rope to each end of the hammock. Two were longer, to reach up to the shoulders, the third adjustable to wrap around the waist. Carefully, they lifted Juan, who was unconscious but groaning softly, into the hammock. Then they tied the hammock securely around his chest and arms, his thighs and shins, almost like a cocoon. They secured his head with strips of canvas.

"It is not so good," said Julio.

"It will do."

"Can two men do it alone?"

"No," said Beto thoughtfully. "They cannot. But they must."

Two of the largest men were called forward, and gently, the canvas hammock was fastened to their shoulders and the steadying ropes tied around their waists. Beto looked about him in the dark. "Who will climb free by their sides?" he asked. The boy with the mockingbird whistle stepped forward.

"You can do this thing, Juancito? There are no extra ropes."

"He is my cousin," answered Juancito.

"Cuidálo," said Beto.

"I will be careful."

Beto nodded and looked again at the group.

A tall figure moved forward in the dark. Memo had not seen him before. As sharp and thin as a skeleton, the man slowly unhitched his pack and laid it on the ground. "Will I do?"

Beto stared at him, as if surprised. "Ah," he sighed. "You have found us. That is good."

"Will I do well enough?" the man repeated.

"There are others," said Beto.

"I can do it."

"You can."

"Then I will go."

"Go carefully, García." Beto touched him gently on the shoulder and turned away. Then he stepped to Juancito and touched him as well, and finally he touched the two bulky carriers, who nodded in reply. "Carefully," he whispered.

Memo watched breathlessly as the men made the first half of the climb and arrived at the spot where Juan had fallen. Slowly, the men passed the small protrusion, easing their bodies upward. Then the torn clouds, which had allowed the moonlight to faintly illuminate the scene above, suddenly sealed themselves over the face of the moon, and little could be seen, though the men strained their eyes against the darkness. After many minutes, they heard muted laughter from the top of the incline. The men relaxed and expelled their pent-up breath in one whispering sigh.

Memo laughed with them. It had been a harrowing vigil, and he felt both exhausted and elated at the same time. He noticed he was rank with sweat, in spite of the fact he had done nothing but watch.

"Memo." Beto appeared at his side. "You'll have to go with them."

"Why?" Memo was astounded at the thought.

"You must lead them down the mountain on the other side. I cannot go with them. It will take most of the night, more than one trip to bring this equipment to the top, so I will have to stay and guide these men to camp when it is finished."

"But I don't know the way. What if I get lost?"

"Memo, it is very simple. Keep the sun on your left shoulder."

"But, man, there is no sun!"

"Then keep the moon on your right shoulder."

"There is no moon either."

"It is only behind the clouds. It will be back."

"But why?" Memo's voice had risen sharply, but he caught himself and spoke more quietly. "I don't know the way. They'll find it just as well without me."

"Even so," replied Beto patiently, "you must be with them on the other side. You must take them to find help in the village. You are known there."

"One of my boys can take them and I will climb this hill in the morning. Only a goat could climb this hill, and I have never been known as a cabrón."

"Memo," the other man said, still patient. "You are the leader, no?"

"Sure, hombre."

"Then you must go. What will your boys think if they know that their leader was afraid to climb a little hill?"

"I saw that man fall," protested Memo, desperately sweating. "It is not a little hill."

"True, he fell, but he had no rope. You will have a rope."

"Sure, amigo, but what if they drop it? Those at the top are only small boys."

"Don't worry, Memo," said Beto, in all seriousness

so far as Memo could tell, clapping him on the shoulder in comradely fashion. "If they drop it, you will fall. It is simple."

"You see, that is what I am worrying about," Memo said earnestly. "And if I fall, I might die. What about that?"

"Then you can be proud, my friend," whispered Beto, "for you will have died for the revolution—for the people."

"Ayeeiii . . ." Memo gestured helplessly. "This is no way to fight a war."

Beto chuckled. "I have often thought the same myself."

So in the end Memo went up the rope to the top of the hill and stumbled through the dark, rocky hillside, taking the lead, which made him catch every obstruction, take every fall, encounter every branch of thorns in his face, and curse loudly and all alone in the night. He no longer cared what they thought.

Finally, in the gray darkness before morning, they reached Memo's still-sleeping camp. He wanted to stop, but the bulky carriers menaced him, so he only roused one boy to tell the others and stumbled on down the path toward the village. Never had he ached so much in his life. Every step was a sharp knife that cut through him. Each breath was a gasp, and his head ached. He had given up cursing aloud because it used too much breath, but he muttered in his mind. He had come to hate these men on the trek down the mountainside. *They are unnatural,* he decided. *They are not men at all. Men do not behave like this. They know when it is sensible to stop and rest, to take a little chicha. These men are not real. They are supermen, maybe possessed by a demonio. Maybe captives of the vampire. Perhaps that is why they have such strength.* But that made him shudder and he hurried faster. *Oh, por los santos, I hope they are not vampires!*

The cocks were crowing and a few old women were stirring by the time they reached the village. Of course the old women stared and clacked their gums together, but Memo was too tired and mean to be embarrassed. He frowned at them and hissed, taking out his hatred for the supermen that followed on the harmless crones, who ducked their heads as if they had been struck and popped back inside the shade of their casas.

Just as he'd expected, when they arrived at the church the courtyard gates were closed and bolted from within.

"Padre!" he yelled. "Father Herrera! Wake up! Get up I tell you! There's a man hurt here." But no one came. A dog barked from behind the gate and that made Memo furious. He kicked at it with his boot. "Lazy old man, get up!" he screamed, and then, unstrapping his gun, he began pounding the doors with its butt. "I'll break your doors down and set your house on fire if you don't get up!" he yelled.

"Shssst! Calláte!" Rita called from behind the door. "What do you want out there? What's all the racket?"

"Open up, woman," he yelled even louder. He was so tired he thought he would kill the next man who said he must take another step. It was time for breakfast and drink and a nap and pulling off his damned wet boots. "Open this damned door in the name of the revolution!" he screamed and began pounding with his gun again.

Slowly, the bolt could be heard grating through the latches. Rita opened the door a tiny crack and peered out at them. "What have you got there?" she asked.

"Move over, woman. Get out of the way." Memo pushed in through the double gates, scowling and grinding his teeth. "We've got a wounded man here. Where do you want me to take him?

But Rita stood mute, unable to answer.

"If you don't answer me, woman, I'll put him in the church! Do you want him to bleed all over your precious church?"

"Is he hurt?"

"He's dying, you fool woman! Where shall I put him?"

"Oh dear, oh my . . . put him in my bed."

"Rats sleep there," spat Memo. "Where's the padre's room? We'll put him in the father's bed."

"No! Oh no, he's asleep. You can't do that!"

"Where is it?"

"Wait. No. I'll need hot water, lots of it, and herbs, no, bring him to the kitchen, put him in there so I can work on him." She pointed in the direction of the kitchen, but Memo was not satisfied. "This man needs a bed," he snapped. "He has been fighting for the revolution. He has to have better than that." Behind him the men stood silently, their blackened faces showing no sign of emotion. The tall one, García, had disappeared.

"You can have mine," she said, looking at the wounded man for the first time. "Ayeeiii, Dios mío, he is hurt."

"What did I tell you, fool!"

"In here, hurry. Please hurry."

"The bed," Memo insisted.

"Over there, my room is in the corner. That way, see. Go fetch it, but do not wake the children."

"What do I care for children?" Memo snarled.

"They will cry. You will care less for that," she snapped. She was already moving for the kitchen, the men close behind her. She cleared a table and wiped it quickly with a rag.

"Place him here," she said, bending over the grimy, bleeding old man. "Oh, el pobrecito. Is he alive?"

The young one named Pablito seemed to take command now. "He will not die, if you move quickly. We will need cobwebs, leaves, clean bandaging, a good strong broth . . ."

"I know what to do, hijo," she said. "Untie him at once."

The men removed the ropes that bound Juan to the canvas stretcher. He groaned as they pulled them from beneath his body.

"His boots," Rita directed, so they bent to remove his boots. "Where else is he wounded?" she demanded, staring at the torn face.

"Inside perhaps," responded Pablito. "I don't know."

"Has he bled from his mouth?" Rita asked.

"In the darkness, lady, we could not see."

Rita touched the old man's forehead. It was hot. Beneath the bandages, the blood still oozed from the slit temple and gashed face. The front of his shirt was red with blood and his face was smeared with blood and burnt cork and greasy sweat and dirt. The whiskers around his lips were brown with blood, and a trail of blood came from one nostril.

"I will need help," she said vaguely. The sight of the old man reminded her of Manuel, for though she had not seen his broken body, she had imagined it. She turned and began throwing bits of wood in the fireplace. Mechanically, she dipped water from a large crock in the corner and poured it into a black iron pot that hung over the fire. "I need help . . ." she mumbled.

"Is there no doctor in the village, lady?" asked Pablito.

"I am as good as any," she answered. "Here in Pasaquina, people are born, and people die. In between they do the best they can. This village has no money for a doctor, señor."

"Do you have medicines—quinine, sulfur, morphine—anything like that?"

"Sometimes we do," she replied.

"And now?"

She shook her head. "Very little, I think. There is not much call for it. Sometimes in the spring the government sends a doctor through to vaccinate the children, and he leaves medicines at the farmacia when he comes, but he did not come this year."

"Where is the farmacia?" Pablito asked.

"Beyond those doors, on the square to the south, but they are not open yet."

Pepito turned to the first carrier. "Open the farmacia," he told the man. "Bring me what you can."

"Señor," Rita protested, "medicine is expensive. Can you pay?" It was not her business to allocate funds for the sick, but she was torn by compassion and was considering waking Father Herrera.

The boy looked at her for a moment, sensing her concern. "We can get what we need, señora." His voice was flat and dull from long experience, though he had yet to achieve his eighteenth birthday.

"Está bien," she nodded. "Then let us begin."

Carefully Rita unwrapped the bandages from Juan's torn head. As soon as the pressure on his skull was relieved he began to bleed again. The wound was jagged and nearly four inches in length. She thought by looking at it he had been torn by a wolf. She could not determine the depth of the wound because the bleeding was profuse, so she took clean cloths, drenched them in hot water, not waiting for it to boil and began cleaning the tear in his scalp.

At first it appeared that the flesh was swollen and debris-filled, but as she carefully removed the bits of dirt, rock and leaf, she found other particles that were at first indistinguishable from the rest. Passing the particles of debris through her fingers, Rita realized that the man's skull had fractured and bits of bone, tiny as splinters of glass, lay in her fingers. She did not know if such a wound could heal or if the man lying beneath her fingers was already dead and merely breathing out of habit, but she grew suddenly afraid and knew they must hurry.

Quickly she rushed to the caldron. The water was hot but not boiling. She dipped up half a pitcher full and then leaned over the unconscious figure on the table.

"I'm sorry, viejo," she said. She parted the wound

with her fingers and poured the water deep into the crevice, washing it and shuddering at the results.

The tall García, filthy and grim in the early morning, seemed to appear from nowhere, so silently had he entered the small kitchen.

"I need help . . ." Rita sighed and seemed about to faint. García reached out and grasped her shoulders.

"Be strong," he whispered.

"I am not a surgeon," she replied. "I have only delivered children and nursed fevers."

"We will need a disinfectant," said García. "Have you some alcohol?"

She shook her head.

"No chaparro?"

"Agua ardiente."

"Where is it?"

Automatically she moved across the room, her knees trembling, and passed behind the faded curtain that hung over the pantry. When she returned she tried to hand the bottle to García, but he shook his head at her. "No, I will hold apart the wound. You pour it."

Rita uncorked the bottle. Her stomach was trembling with her hands now, and her face had broken out in a sweat, though the morning was still cool. Gently she dripped the brown liquid into the wound.

"More," commanded García. "Flush it clean."

She wondered if he knew about the bone splinters.

García was not satisfied until the bottle was emptied. "Good," he said. "You are doing well. Now we must stitch him. We cannot heal this wound with cobwebs. Go get your sharpest needle and your strongest thread."

"I cannot," she faltered. "I can't do it."

"I will tell you how," he said quietly. "I will help you."

"Are you a doctor?"

"No, señora, I have never been to school."

"Then how do you know what to do?"

"I was in the army once."

"Then you were a doctor in the army?"

"No, señora, but in the army there are never enough doctors, so I have held together many broken men."

"¡Qué terrible!"

"Hurry, señora, we have no time. Hurry."

Quickly she was back with a needle and thread. García examined the needle and thread, and though he was not pleased, he knew it was the best she had, so he accepted it with a nod. First he made her hold the needle to a flame until it was clean and then he told her to double the thread. When she had finished, he tested it for strength, then spit on his fingers and ran them down the thread, twisting it together to make one strand.

"Now I will hold the man," he said. "It is very simple. Just think you are making a quilt and you want to make it strong."

He directed her fingers to the top of the gash and said, "Begin here. Push the needle through the flesh—do not be shy—the quicker you go the less pain he will feel. Push the needle through this side, here, then out here again. Quickly now, woman."

Taking a deep breath and fighting not to close her eyes, Rita obeyed. Her stomach rebelled and rose up into her heart.

"There. Good," said García. "Now you have made the first stitch. Pull it up tight; pull the threads together now and tie them. Quickly."

So Rita tied the first knot and bit off the thread with her teeth. The warm blood smell filled her nostrils and made her head swim.

"Good," said García. "You are a good woman. That was fine. Now do it again, just here. Begin quickly. Do it again."

The silver needle plunged into the fat, swollen flesh, which resisted with a force she had not expected. One more stitch. One more knot. One more bite with the sharp teeth. Red flecks swam in front of her eyes, but García kept talking.

"Again, just here. Be quick. Do not falter. You are a good woman. You are doing fine. That was a good stitch, a good knot, a good woman, a fine stitch, a fine knot, a fine bite, a fine woman . . ." until somehow she had sewn together the ragged wound as tightly as the seams of a fine dress, until somehow she had sewn it so tight and fine that blood no longer flowed but merely oozed softly.

"Now bring the water," said García. "We must clean him again."

She fetched the clean cloth to swab him but García said no, so once again she poured the hot water across the wound, while the pitiful gray-faced figure beneath her groaned and opened his eyes to roll them unseeing into space. By now Rita's hands were shaking hard, as if she had palsy.

"Now we must have a salve, an ointment. What have you got?"

She only shook her head in dismay.

"Is there no calamina plant here in the courtyard?"

For a moment she didn't seem to know.

"The cactus you use for burns," he insisted.

Then she nodded. "In the back. In the alleyway."

"Fetch it."

Soon she was back with a thick green spine. Now she knew what to do without being told. She slit the plant up the middle with a kitchen knife to expose its moist, pulpy interior. Then she cut a slab of the green, jelly-like pulp and laid it against the clean wound. García held it in place while she stripped cloth, and then she laid the strips delicately across the wound. When she had formed a pad, she took the longer strips and began wrapping Juan's head.

"No. More. Tighter," instructed García. "It must be held tightly, do not be shy." So Rita bound the wound tightly with the long white strips of cloth.

When Juan's head and face were clean and bandaged, García began to strip him.

Rita gasped and drew back.

"Not now," said García, "do not desert me now, woman. It is not yet time for fainting, and besides, have you not seen a man's body before?"

She shook her head. "Not like this."

"Then this must be the right time," he said with a grimace on his lips which could have been either the beginnings of a smile or an indication that he was clenching his jaw.

So Rita helped him strip the man, stood by while he examined the body as best he could and waited for his conclusions.

García ran his hands lightly but insistently over the body, which appeared as wilted as sun-baked celery, and then drew back and sighed. "I think he has broken some ribs," he said to Rita, whom he treated with the respect due an equal. "Pray to your god, woman, that his ribs have not punctured his lungs. Pray that he does not choke to death on his own blood."

Then he grinned at her. "You must do it, for I am not a praying man. But I am afraid that praying is all we can do for this problem."

So, carefully and gently, they washed Juan in tepid water, wrapped his torso in a clean sheet, and laid him gently on the bed which had been brought in from Rita's room. Though it was early summer in the mountains, García covered him with three blankets before he was finished. Then he turned to Rita with a sad smile and said, "We have done all we can do, and now it is time for your prayers, señora."

Tears came into her eyes, and she pressed her blood-smeared hands to her mouth.

Mistaking her anxiety for remorse, García asked, "What's the matter? Don't you pray either?"

For the first time that morning, Rita began to cry. She pressed her chubby hands tighter against her face, shook her head mutely, ran out of the kitchen, across the courtyard, fell to her knees in the beaten earth and retched out the green bile that had risen to her throat. She retched and sobbed intermittently until García squatted beside her in the dust and put his long, thin arms around her shoulders.

"Poor girl," he whispered, "don't cry. Don't cry,

please. Please. Stop crying and I will pray with you. I promise."

Finally, Rita sat back on her haunches, wiped her
mouth on her sleeve and looked sadly at García.

"What are you?" she asked. "If you are not a doctor,
what are you?"

"A guerrilla," he said. "A revolucionario."

"No. It is more than that. I have seen soldiers before, and you are not like them. If you fight, you are still
not a soldier, not in your heart. I know that."

She waited, but he only dropped his eyes.

"What are you?"

He sighed and turned his head. He seemed to be
staring at the purple bougainvillea that grew up the back
wall of the courtyard, or perhaps the colored chickens that
scratched in the dust, or the old red dog asleep in the
shade.

"A poet," he answered. "I am a poet."

She relaxed then, understanding what she had
seen.

"You are a writer of songs," she mused, as if it had
not occurred to her before just where songs originated.

He nodded, still looking at the bougainvillea. "I
write songs that make their music without the strums of the
guitar or the mandolin. I write songs whose notes are
words, whose melody is the language of my people."

"I would like to hear your songs," she said.

"It is the sadness of my songs," he said softly, "that
the people whose language makes their sound never hear
them, never hear the rhythm of their own voices. Only a
few in the cities can hear my songs, and they do not under-

stand them, for their voices are different. Their voices are the rattling of the streetcar, the clicking of typewriters, the whine of old trucks, and the pounding of machines. Their sounds are hard and have much monotony. So when they hear my songs, they cannot tell they are listening to music, but believe they are only reading the words of a man with sparse vocabulary."

Rita sat very still, facing him. His eyes were gazing in her direction, yet he looked not at her, but through her, toward some point far away. After a moment, his voice continued.

"When you speak, you make the sounds of the red bougainvillea; you make the sounds of the mist that lies on the grass in the mornings, so that to the eye it appears not green but turquoise. Yours are the sounds of honey melting, and satin-white hibiscus. The sounds of my people are the sound the wind makes when it is still, the sound of lightning that is yellow on a black night, the sound of an infant's skin; these are the sounds you speak, the sounds I write."

"I think I can hear them," she said.

"Only the Latin can make such sounds with his language," he explained. "Though vision and beauty may exist in many other races, it will not be found in their words. Maybe music will be seen in their paintings, their tapestries, their coastlines, but it will not be sounded in their words. Listen. I'll tell you.

"Mi corazón . . ." he whispered. Chills rose on her arms at the sound of it. "It soothes the ears. But in inglés . . . my heart—it bites you."

"My *hart* . . ."

"Mi corazón . . ."

"Querida mía . . ." she laughed. "That one dances!"

"My beloved—it strikes you!"

She laughed again. "Be-luv-ed—it hits you!"

"Novio mío . . . how it caresses. But in inglés, listen now—my sweetheart."

"Oh no!" she cried, clapping her hands to her ears. "Sweet-hart," she repeated, biting the "t's." "No. I don't like it. 'Hart,' it is like a stick to me."

He laughed. "To be sure, as I told you, only a Latino can make real music with his words. That is something all the science in the world cannot give the gringos."

"Then we are very rich," she said gaily. "We have a prize no one can steal."

They laughed together and for a moment forgot the blood on their blouses, the dirt and cork and sweat that smeared them.

He smiled at her and she smiled in return. "You are a good woman," he said. "What is your name?"

"I am Rita," she answered. "And you are El Ave Cantora, the Songbird."

"Yes," he laughed, "from now on that is my name—Ave Cantora."

Then her raised eyes twinkled. "But you look like a lizard!" she laughed. "Go and wash yourself and I will make coffee and buy us pan dulce."

"Thank you," he said.

"Never mind," she teased, rising lightly, "you will pay. In exchange you will sing me a song of words."

"A fair bargain."

She walked away toward her room to change her chemise. At the doorway she glanced quickly over her shoulder. He was still standing in the courtyard looking at the bougainvillea.

Eleven

When Sister Magdalena came out of her chapel room and saw the big, filthy men on the veranda, she was at first startled, then frightened. But within a moment her fear was replaced by excitement. She looked at them avidly out of the corner of her eye, making no move to acknowledge them, and they, in turn, resolutely ignored her. When she had washed, she headed back across the sun-drenched courtyard toward the kitchen in search of Rita, eager for news of these strangers.

What she found instead of Rita was a dying old man, grisly as a mummy in his bandages and blankets. His blank, unconscious eyes fluttered and stared at the sound of her entrance, causing her to think at first that he was awake, but when she asked him in a flustered small voice to forgive her for disturbing him, he only stared more and groaned. Sister Magdalena had not previously had the opportunity to study a man closely, so she stared back. As she watched, Juan's body shifted on the cot and he cried out in pain. His eyes rolled back into his head and then vanished behind the closing lids.

"Ayeeiii, por Dios . . ." she whispered, realizing the man was not conscious, and moved closer to his side.

Transfixed, she stared down at him, studying the lined face, the bristled chin, the hollow eyes, the parched lips. Then, tentatively, she reached out and touched him. Expecting to recoil in horror, she instead found herself petrified by warmth and compassion. A deep ache was in her heart and tears came to her eyes, tears of love so intense that she thought she had never before this moment known the real feeling of love.

Confused and embarrassed by her reaction, she stepped back and tried to leave the room, but the sight of the dying man held her immobile. He was old. He was ugly. He smelled rank in the warm morning. Why did she feel this enchantment? Never before had she responded to a human being with such instant heartache and desire. She wanted to wrap her arms around his thin, dry frame, and soothe his black hair with her fingers. Was it a mother's love? It did not seem so. She stepped further back from the suffering figure, as if to see more clearly what she herself was feeling.

"It is the love of Christ," she said aloud. "It is the love a true Christian must feel for his brothers."

But something came in front of her eyes, something she could not see clearly; it was only a mist of moving shadows. She wiped her hand in front of her eyes to make it go away.

"I have a Christian feeling for this poor man," she insisted to herself. "Maybe I am going to be a real nun after all."

The vision came back, placing itself between her and the silent form on the cot. There were shapes of human bodies, but the images would not be still. Gray and undulating like liquid in a jar, they floated before her.

"What am I trying to see?" They were forming themselves, taking shape. Standing people, kneeling people —and the sounds. She could hear sounds, too.

"I don't want to see it. No. I don't want to see it." She wiped her hands across her face. "Don't make me see it!"

She fled from the room, saying again, "It is the love of a Christian for another Christian. It is God's pure love. I am becoming a nun. I am learning to be like God. I am feeling only God's love. It is because he is sick. No, he is dying. God would love him. I don't want to see it! Let God love him, not me. I can't see it . . .!" And she ran down the veranda, not seeing the men leaning against their posts with their coffee mugs and their automatic rifles and their black combat boots.

"Father," she was calling before she reached his door, "Father, come quickly. Wake up, come quickly, Father . . ." She pounded on his door, and heard the old man grumble in reply. "Wake up, hurry. Wake up, Father. There is someone dying. Come quickly, he needs you. Please."

"Hush, woman!" A dark-faced stranger had her by the shoulder and was shaking her.

"Please, Father . . ." she continued, not realizing that she was hysterical, or why. Her voice only grew louder.

"The man is not a Catholic," the big one said. "He does not need this. Keep quiet."

But she was still crying and calling and pounding on the door. From inside, Father Herrera could be heard mumbling and shuffling about the room.

Suddenly, the big one turned her around and

slapped her hard on the face. "¡Calláte!" he commanded. "If he is not dying, you will kill him with your noise. If he is, have the decency to let him die in peace."

The commotion had roused García, who had been resting on a pallet in Rita's room while she went for the pan dulce. He came striding across the courtyard with a frown on his face. "Take it easy, muchacho," he told the big one. Gently he lifted the thick hand from Sister Magdalena's shoulder. "She's only a child. Here, go back to your coffee. I'll take her."

He guided the girl away from the men and sat her down on a bench beside the well near the back door of the church. A small blooming tree grew between the well and the front wall of the courtyard, and though its bloom had vanished, a few yellowed petals lay in the dust beneath the bench, their smell still faintly pleasant. García held her shoulders, bending over and looking deeply into her eyes. She was still crying, but she was inarticulate now.

"Señorita," he said, "will you wait for me? I will bring you a cup of coffee if you will wait for me. Can I leave you? Will you stay here for me?"

She nodded but did not look at him.

Momentarily, he returned to the bench with the coffee and seated himself beside her. But he did not intrude by allowing his body to brush hers or by looking at her face. He sat with his long fingers laced around his knees, looking at the bougainvillea on the back wall, and when he heard her make sipping sounds, he began to talk quietly to her.

"The old man's name is Juan," he said. "He is a very good man—a kind man. Before the war, he worked in the fields as a trabajador. He lived in a small village at the base of another mountain, far to the north of here. But the mountain was very like your own, gentle and good, with rainfall always in the proper season. He had a wife, a daughter, and many sons. They were very poor people; he is not an educated man. But he had a good family and much to be thankful for, señorita.

"His daughter had a baby girl, so Juan even became a grandfather. He was very proud." García smiled as if he had been there to see the baby's christening, as if he were about to say, "I remember it well." But he listened for a moment, and hearing that her breath was steady and even, went on.

"I have known him for a long time. I have known Juan since I first came to join the revolution, and I have heard many stories about his home and his village and his family. He has been a good man all his life, señorita, and I can say that with conviction, because in the time I have known him, I have seen him do much good. He is a man with a generous heart, and he has had a good life." He paused and rubbed his fingertips together.

"If he is to die, señorita, it will not be a tragedy. Do you understand? He has had a good life, and not a short one for these times."

Sister Magdalena had recovered herself somewhat. "I'm sorry. I don't know what happened to me."

"You need not be sorry. It was natural for you to go for the priest, you are a sister of the church. But you need not be sorry for Juan either, señorita, for he has nothing to fear. And he has no need of a priest."

She dropped her head and seemed about to cry again.

128

"If there is some God, as you believe," he continued softly, "then he has made us all and will surely understand our natures. If your God exists, señorita, then he must be at least as good a man as Juan, and Juan, I know, would never turn away a man of sincerity."

"It is not that," she said, looking at him for the first time. His face was clean now, and its thin, pale shape seemed innocent. His brown eyes were large and sad, and his mouth wide and sensual, though his lips were narrow. He had the look of a tired man who had resigned himself to unspeakable fates, and yet at the same time, she could see that a contradiction was present.

"It's not that," she said again, collecting her thoughts, glad of his company. "I can't explain it, señor. I've made a fool of myself, for I am not a believer either. I am like you and believe only in the people, in the revolution. I am so ashamed that I have made a spectacle of myself."

"Don't be ashamed, hija, it is natural enough."

"But I am not a real nun. I have only come to this because I was homeless and had nowhere else to go. If I had been a boy I would have joined the guerrillas and fought; but as it is, I could only accept what was done to me. You can understand?"

"Of course."

"I don't know what frightened me so. It was foolish to want the priest and confession, and all the rest of it, but I did. I wanted the Father in the room with me and that man, just terribly, I did. Now I am so ashamed."

"There's no need. Perhaps you have not seen very much death. Death does strange things to us all."

"I have seen it," she whispered. "But it was long ago. Maybe if I had seen enough, like you, I would have become used to it."

He shook his head. "No, hija, you would not."

Father Herrera came banging out of his door and stopped short on the veranda. Looking about at the strangers, his mouth fell open, and his hands reached to the tie around his tummy and fumbled with it. He had meant to reach for his crucifix, but in his distraction, had pulled at his belt instead. "Oh my . . ." he whispered. His nervous hands felt his plump cheeks and pulled at his chin, and his little eyes darted quickly from one face to the next, fearfully. "Oh my." He spied Sister Magdalena on the bench by the well and shuffled in her direction with as much authority as he could muster.

"What's happening here?" he demanded. "Who are these men?" He spoke to the girl, carefully avoiding the man at her side. "Are you responsible for this?"

She stood quickly to address him, her long training overcoming the rebellious spirit she had just been expressing. "Father, I don't know how they came, but they have brought a wounded man, and, Father, I think he is dying."

"What, dying?"

"In the kitchen, Father. He is dying in the kitchen."

"Oh my, God save us, in the kitchen?"

"Yes, on a bed in the kitchen, and he's been hurt badly."

"I know that," Father Herrera snapped. "How could he die if he weren't hurt badly? Where's Rita?"

"I don't know, Father."

"Please," said García, stepping forward, "excuse

me, I am García. We have brought this man from over the mountain because he was injured badly and in need of medical attention and rest."

"But I am not a doctor!" protested the Father, and then quickly added, "Can he make his confession?"

"Your criada has made coffee, señor, and gone for pan dulce. She should be returning momentarily."

"Rita, gone to the market?" said Father Herrera as if it were impossible, and in fact it was. The thought of an emergency without Rita was unthinkable. "Quick, child, go and fetch her," he said to Sister Magdalena.

"She will be back shortly," interceded García, hoping to calm the flustered little man. "May I bring you a cup of coffee?"

"Coffee? Yes, coffee. That would be good. Coffee."

"Would you like to sit here, señor?" García offered him his place on the bench.

"Yes. I will sit here," responded the priest, sweating and feeling ill in his stomach again. *Oh, where is Rita?* he thought to himself. *Why does that woman desert me?* But he sat on the bench and laced his stubby fingers across his stomach, trying to appear calm.

"I will help you, Señor García," volunteered Sister Magdalena. She did not want to be left alone with the priest, who might still decide to hold her responsible for the rabble intruding themselves into his courtyard.

So Father Herrera sat alone on the bench and blinked.

"What shall I do?" Sister Magdalena asked García as they poured coffee and measured out the brown sugar. "He will regain his wits and insist on offering a blessing for

Juan." Using the stranger's name made her feel a sense of belonging, as if she were one of them, part of the group of the hard, brave men.

"Never mind," he said. "Let him perform his duties. What harm can a few words do a dying man?"

"But what about Juan? Would he want this?"

"Who can say what a man will want when he knows he is dying?"

"Will he die?" she asked, glancing at the still figure. Juan's breathing was softer, but still rasping in his throat. A trickle of blood had appeared at the corner of his mouth.

"Wait and see," said García. "That is all we can do." But in his heart he knew the answer. Juan would die. Juan would die and be buried and his grave would be marked only with a rough stone on which his name was scratched. He would die, and if there were any left behind in the village that had been his home, the chances were they would never know the manner of his death or the day when it had occurred. They would only wait, if there were any of his family left, and after enough years had passed, they would assume he was dead because it had been so long, but they would never know, so from time to time, even after the passage of long years, they would wonder . . . if they were not already dead themselves. That is what García knew but did not say.

"I hope he will not die," she said softly.

"Yes, that is what you must do," he agreed. "Hope."

"And I hope Father Herrera will let you stay." She ventured a glance at his face, which she thought was like the face of a sad horse. And then she remembered the stranger in the church with Memo. His was a beautiful face. All the fighters for the people should have such faces, but they did not. The old man dying on the cot had the face of a drying banana leaf, and the one who had shaken her in the courtyard had the face of a bear.

García smiled at her. "I will speak with the padre," he said. "You take his coffee, and I will come shortly."

When she left the kitchen, he stood for a moment, looking down at his friend and listening, as if the sound of the man's shallow breathing could tell the hour of his death. But the sounds only repeated themselves and made no music for the Ave Cantora, so he turned from them, and went outside to speak to the men. *There is no good music in dying,* he thought. *It makes no sound I like to hear. Of all the things in life that matter, the two that are most important cannot be said.* And he thought of birth and death and how little was in between, as he made his way down the veranda to squat beside the other men.


When Rita returned with the pan dulce, she discovered that Father Herrera had withdrawn himself to the church to say his morning prayers. He had taken Sister Magdalena with him to bolster his quavering sense of security, and was making an especially long business of it, since the sister had reminded him that he should say special prayers for the dying man. Father Herrera distinctly wanted the man not to die in his kitchen, for he was sure it would foul the air, so he prayed more fervently than usual that morning, and stayed conveniently out of sight until Rita was ready for him.

Rita had been looking forward to a private breakfast with her new friend, but the discovery that Father Herrera and Sister Magdalena had both emerged for the day complicated her position. The last thing she wanted was for the girl to think she approved of revolutionaries. If she were discovered in private conversation and private breakfasting with García, she might not be able to explain that he was different, not a revolutionary at all. She pondered the matter as she placed the pan dulce on a tray. The man behind her on the cot bothered her not at all. She had done her work, and done it well, according to García. The father was saying his prayers, and so the pobre would recover in due course, or he would not. It seemed obvious to her.

García solved her problem for her. He stood in the doorway behind her. "I have spoken to the men," he said, and when she turned to look at him, she saw that his eyes were empty. The Ave Cantora had flown, leaving behind only a forlorn, exhausted stranger. It was not the time for a song of words.

Rita handed him the tray. "Here," she said, for she was instinctively a mother as well, "the men are hungry. Give them these sweet breads to eat. I will bring you a pitcher of milk."

"Thank you."

"You can tell me after you have eaten."

"We cannot leave just yet," he said.

"I thought so." Her back was toward him so he didn't see the hurt in her eyes.

"We must wait here a little while for instructions, but I hope we will not trouble you for long."

"There will be no problem," she said, still not allowing him to see her face. "I will see to it."

"You are very kind," he said. A look of regret slipped across his features, but Rita could not see it. "Thank you."

He left with his tray of pan dulce, and Rita kept dipping milk into the pitcher. The Ave Cantora slipped into Manuel and Manuel into him, and they into the others until there was only one soldier in her mind. "You are always leaving," she said. "Leaving is what you do best."

She wiped her face with her skirt and took the milk pitcher to the small group of men sitting quietly in the shade of the veranda. They did not look up at her, for they were no longer curious, and she did not look at them. She did not want to see them. She walked slowly across the courtyard and entered the church by the back door. Father Herrera and Sister Magdalena were kneeling before the crucifix, he on the raised platform before the altar, and she beneath in the aisle. Rita crossed herself and slid behind the first pew. She bowed her head and held her hands together in an attitude of prayer. She closed her eyes while Father Herrera's soft, monotonous voice said the words which meant nothing but sounded like a lullaby. Behind her eyelids in the dark, Rita saw only the tall, thin body of the Ave Cantora lying naked in her bed. She saw only her plump hands stroking his leanness, running down his chest across the bones of his ribs where they came together in front, and down his long thighs, so tight and hard. In the darkness behind her eyes, she saw herself leaning forward over him, her lips touching his thin lips which were still sweet with songs, and her full breasts invoking his smooth chest, while her hand ran down to stroke him between his legs. She felt neither shame nor guilt nor impropriety. She felt only the longing which would not be satisfied. *You will be leaving,* she thought. *That is all they have taught you in this war. All you have learned is leaving.* And she cried silently into her plump, empty hands.

)

# Twelve 🌙

Beto arrived in the village at siesta time. The supplies had, after much labor, been brought over the mountain and were stacked temporarily in the camp of los muchachos. There the rest of Beto's men rested and waited for Beto to investigate the situation in the village.

Entering the church courtyard stealthily, Beto stood in the shadow of the big gates to observe. What he saw was only stillness, the exhausted aftermath of the previous night's crisis. Memo had fallen asleep early that morning, had waked only to take lunch, and had moved into a storage room to sleep again, but the other men had not been so fortunate or so relaxed, and they dozed with their rifles across their knees along the walls of the veranda.

Father Herrera had returned to his room after lunch and bolted his door. He had no intention of emerging before the evening meal, since he found himself in a state of enormous confusion regarding his visitors. The larger part of him wanted to turn them out with a kick and a curse, while the smaller voice in his head said he should welcome them with open arms and be grateful that God had sent him a mission. But the surly, staring men did not seem interested in making confessions or saying prayers; he was certain they wanted only his food and not his services as a priest. The problem was too great for him to bear alone, but Rita had remained busy throughout the day with feeding the men and hovering over the dying old one in the kitchen, and now, exhausted after the morning's work, was sleeping on an improvised pallet in her shady room. So the courtyard was still.

Only Sister Magdalena was awake. She had tried to pray in the morning with the padre, but the excitement of the visitors from the revolution had made her prayers feel hollow in her mouth. She had tried to pray over Juan's inert body, concentrating on each word separately, but their meaning eluded her. She had stood by while Father Herrera gave the last rites, but had found no magic in the sound of it. All she could feel was annoyance that the flies settled on the man's bandages and she couldn't move to shoo them away. She had served food, run errands, attempted conversation with the quiet men, and finally given up on all of it. The confusion of the early morning remained to haunt her. At one moment, she wanted desperately the consolation of the Church, yet the next, she wanted nothing more than to be free of its bonds. She was split in half and standing in two worlds.

At first she delivered herself a diligent lecture on the foolishness of all religious teachings in general and Catholic edicts in particular, which felt good while the reviling lasted, but the effect seemed to sputter and die when she began to think of what one might believe instead. She was little educated, except in the curious mysteries and inexplicable decrees of the Catholic Church. She had no foundation in logic; indeed, it was foreign to her nature. But even had it not been, it would have been of no help anyway. So she floundered. When she thought about the church, she thought principally about the convent, where the sisters had been piously cruel, and she was often hungry and cried in the night. Alone. She had been alone. Though she told herself she had not been hungry, it only meant that she had not starved. And when the sisters had said she was evil for feeling lonely in the presence of Christ and the Virgin, she had felt it, the wickedness they told her was there. Wicked and alone.

Remembering all that, it was easy to believe, as many of the revolutionaries said, that the church was wrong because it meant to keep the people slaves to itself and to the government. But she knew nothing about the government, and still less about its relationship to the church. So the result of her deliberations on this point was little more than a vague swirl of ardent feelings for the revolution which would set the people free.

But before lunchtime, she had reversed herself, and decided that perhaps the reason the revolutionaries hadn't yet won the war was a lack of earnest praying. Though it was clear to her that the two worlds didn't mesh, the one being composed of seemingly mindless violence, and the other of seemingly meaningless words, the idea of conjoining them encouraged her enough to partake of the noon meal.

She had meant to retire to the church to pray immediately after lunch, when she could be sure of privacy, and had, in fact, entered the sanctuary. But once on her knees, she had found herself at a loss. She didn't know what to ask for. Should she say to God, "Let the guerrillas kill more soldiers"? Or should she say, "Dear Father in heaven, let the soldiers from the capital die of their own accord"? Or could she ask, "Sweet Virgin, make the villagers give the guerrillas more money for weapons," knowing as she asked that the villagers hadn't enough money for food, clothing and medicine as it was? There was nothing she could discover to pray for except, "Make Juan well." That was a very short prayer and could only be rephrased in a limited number of ways, so she was shortly finished, and just as quickly frustrated, and, if she had admitted the truth, angry.

That is how Beto found her, sitting beneath the dogwood by the well, pouting. He was drawn to her because he was exhausted and not in his right mind and because her lovely face in a pout was even more beautiful than in tranquility.

She was staring across the courtyard with a hard frown on her face. Her brows were pinched together and her lower lip protruded slightly. It was rosy for she had been biting it in frustration, raking her teeth across it like a child.

"Buenas tardes, señorita." He didn't know why he stood there. It had not been his intention to approach her. She was just a girl and had no authority. He had no business with her. And she was a nun, the kind of woman he despised. The thought of a healthy young girl abdicating into barrenness made him indignant, made him almost violent —but he was not thinking about that as he stood over her small form in the quiet afternoon.

She frowned at him, for a moment not seeing who he was. Then she drew back and lost her frown. Her eyes widened and her soft lips parted. "Good afternoon, señor."

Beto was lost in her face. It was an unusual thing for him, for he did not admire women as a general rule. He used them, but he did not admire them, nor find them a temptation to turn him from his discipline.

"I have come to see the old one," he said finally.

She was watching his eyes. There was a fascination there, a lure, like the glitter of diamonds and at the same time, like the emptiness of death. "I have seen him," she said. "He is sleeping."

Beto glanced around the courtyard.

"He is in the kitchen on a bed. They have stitched his head and bandaged his face, but he has a fever. He has not waked since they brought him this morning."

Beto sighed.

"But I have been praying for him . . ." she added hopefully.

When he looked at her again there was a trace of amusement in his eyes. "Have you?"

She blushed. "It can't hurt, can it?"

"It could, in some ways."

She spoke almost carelessly. "If he is an unbeliever, it might kill him, I suppose. Do you think he is an unbeliever? Have I killed him, señor?"

Then Beto grinned at her. "Please, lady, we revolutionaries are having a hard enough time as it is. Por favor, no more prayers." Then he cocked his head and almost winked at her.

"Say, muchacha, is it you who have been giving us such a hard time? Have you been asking God to save all good Christians and crush the infidels? Is that what you've been doing, eh? Is that why we've been losing this war?'' He shook his head in mock grimness. Sister Magdalena almost opened her mouth to protest, but then she saw that he was playing a game and inviting her to play also.

"And all this time I've been thinking it was our fault," Beto continued teasingly. "Too few men, too few weapons, little things like that—now I see it's been you all along . . . ah, lady, you are a very powerful sister. Maybe when the war is over they'll make you a saint! Unless I can get you to change sides before it's too late for us poor infidels."

Though Sister Magdalena had never played with a man, like most girls she knew instinctively how it was done. "Oh, señor," she said with pretended dismay, "you wish me to desert my mother the Holy Church and pray for victory to go to unholy pagans? How can you ask it of me? Can't you see I'm a very pious girl?"

He leaned against the wall, lighthearted, light-headed, wanting to forget the war. "As a matter of fact, señorita, I can see it very well. I would even go so far as to say that never in my life have I seen such a pure and pious lady. You are by far the most devout Catholic I have ever encountered."

"How can you tell?" she asked, inviting him with her eyes to continue the game.

"Well," he said, most seriously, scratching his head to help him think, ". . . there is something about your nose . . ."

"What!"

"I cannot be sure . . . I hesitate because I wouldn't

like to make a mistake. Be assured it is not my intention to mislead you . . ."

"Tell me what it is about my nose!"

"Let me study it. Here. That way. Turn your face like this . . ." touching her chin, " . . . now, the light is directly on your nose . . . hmmmm . . ."

She giggled and cut her eyes at him. "Can you see it now?"

"Ah," he sighed, "no, I'm sorry, it must not have been the nose. You see, I told you I was not sure. Maybe it is the mouth. Can I take a look at your mouth?"

Her lips opened in a wide smile.

"No. No. Now I have it! Not the mouth. It is the teeth! That's how you can tell. You can always tell a pious woman by her teeth, you see!"

"Tell me about my teeth, señor." She tilted her face further up so that he could see them more plainly.

"Well. Let me see, you have one, two, three, four . . ." he touched her front teeth with his rough brown finger, "and they are white! They are neither red nor purple not green . . . but white!"

"I begin to doubt you, señor," she said, drawing back coyly. "Everyone has white teeth."

"Do they? Is that so? Then perhaps I have made a mistake again. It must be the forehead, do you think so?"

"I think not."

"But your forehead is exceptionally smooth."

"Even girls who are mothers before they are married sometimes have smooth foreheads."

"Ah . . ." He seemed to consider this. "And it is not pious to be a mother?"

She blushed. "I do not know if it is pious," she said, softly, "but I think it must be wonderful."

This astounded him, and so he did not speak for a moment. But she was not through with him. The play had dispelled her morning frustration; in fact, her problem seemed not to exist anymore, because when laughing with him, she was not standing in two worlds, just one.

"But, señor, you have not answered my question. And how can I know if I am pious enough to pray for you, if you do not tell me? I cannot see myself. Only you can tell me."

He smiled again, feeling a little self-conscious, but willing. He had thought he was playing with her as with a child, no more than that, but once he had begun the looking, he could not stop. The girl drew his eyes to her. There was something in her face he had forgotten and needed to find again.

He rubbed his finger against his chin whiskers and pretended to think. "We have very little left to consider, since there is so little of you available to see . . ."

She wondered how the whiskers felt beneath his fingers.

"So, we shall consider the eyes . . . they are large, that is clear, so perhaps they see a great deal more than other eyes. They are dark, as dark as a moonless night, so they have a great mystery in them—and a pious woman is undoubtedly a great mystery—anyone will tell you that."

Again she laughed, to encourage him. She did not want the friendliness to stop. There had been so little of it. That made it precious, dear.

"Let me look at your eyes." He bent closer to

examine her eyes. Automatically she widened them even more and stared back at him with utter somberness.

"No!" he cried, jumping back suddenly. "No. It cannot be the eyes."

"Why not?"

"I can't tell you." He shook his head.

"Why not?"

"It is too serious for a girl."

"You must."

"Oh, no, lady, I can't."

"But what did you see in my eyes?"

"Things you would not like your mother to know . . ."

"But I have no mother . . ."

"Nor I."

"Then we are safe . . . tell me."

"But what if someone overheard? No, I think we should forget the eyes . . . here, let me examine your hand." He took her dainty hand in his rough palm. For an instant both of them were rigid. The shock of their touching sat them bolt upright and they stared again at each other.

But when their eyes met, the electric pain that flowed through their palms turned to a warm, aching sweetness, a liquid flowing back and forth that bound them. He held her hand and spoke again, but his voice was a whisper.

"Let me tell you about your hand instead . . ."

Her soft voice was intent, for the flowing stirred the juices in her groin. "No, first you must tell me about the pictures in my eyes . . ." She hesitated. "And I will tell you about the pictures in yours."

So he let himself be tempted, aroused and hard between his legs, and tight in his chest while hot in the palm that held her hand.

"I am not an artist. It would take an artist to draw such pictures . . ."

But he did speak.

"In your eyes there are horses with flared nostrils and flying manes, running across ridges of rock against a black sky . . . and flying green clouds crossing the moon . . . it is cold, the night is very cold . . ."

And she replied.

"While you have orchids, so purple they are almost black, and furry, I can touch the petals and they feel like a rabbit's down . . ."

He was lost. He resisted, but he was lost.

". . . and red, rivers of red," she went on, entranced, "as red as satin in the church, as red as lava on the mountainside, red, but it cries. It twists, and curls and flows like lava when it is hot, but in the crying there is . . . something . . . I do not know the word for it . . . it is a birthing blood, from the earth's womb, something that is like strength, like power, but much more . . . I haven't seen the word. It is heat, and joy. It is crying . . ."

". . . and mountain pools," he answered her, "pools of still cold water against the skin, and you float there . . . so white . . ."

"I have seen you," she said, ". . . I have seen you broken . . . and then full again. How is it?"

". . . there is an ember glowing, with flames dancing above it and orange light coming out, the smell of balsam resin is in the light, and you are the smell and the light and

146

the heat . . . you are . . ." He felt it. He could not believe it. He could not resist it. "The heat . . ."

"I have seen you before," she said, and he was placing his dry hard lips against her soft ones, and his hand clutched hers tightly. He was leaning over her in the sleeping courtyard, kissing her. His mouth engulfed her and she stroked his lips with her tongue. He touched his tongue to hers and the heat and the orange light flowed like liquid. Then she pressed harder, eating him with her lips, taking in his hard mouth, his rigid tongue.

". . . give me . . ." His rough hand was on her smooth face, walking down the slope of it as lightly as a hummingbird feeds.

". . . in my dreams . . ." she murmured, "I have seen your face."

He wanted to say, "I love you." He felt it to be true, but he was trembling. He was weak. He was a mewling infant, helpless at his mother's breast.

". . . give me . . ." she whispered.

He wanted. Had he ever wanted so? This thing that was between them was not of lust but something more. He wanted and shook from it. "Ah, por Dios . . ." He shuddered and pushed back from her. "Little one, I cannot."

But her large eyes were full of tears, so he knelt in front of her and took both her hands in his.

"I—I apologize," he stammered and thought in a moment he would be ashamed for having tears also. "Please, little one, forgive me . . . I'm sorry, I'm only a man . . . please, forgive me."

The tears rolled down her cheeks. Her voice caught

in her throat and her big eyes frowned. "You are not," she said. "You are not a man . . ."

"I didn't mean to do it, little one."

"You are not a man," she insisted.

"Please forgive me. I promise it will not happen again."

"No," she shook her head, "you cannot promise that . . . I will not let you. I don't even know your name, but I know your face. I have seen you all my life. I have dreamed of you and now you have come. But you are not a man like other men . . ."

"Little one, don't talk like that. You are just a child."

"I am a woman. I have been old enough to bear children for three years."

"You are a bride of Christ, hija, remember that."

"No. I am no one's bride. I have never been a bride."

Beto was perplexed as he had never been before. He wanted her still with such ferocity it felt as if it would break him in two, yet it was not only his body that wanted her, but all his soul as well. But he couldn't have her, even if she were free to be taken. Loving made a man foolish. He could not let himself love, not now. Men in love do not win wars.

He tried to think, but she was crying. He wanted to take the tears from her smooth cheeks with his tongue. He wanted to press her face against his and feel her long eyelashes against his brow. He wanted her nose to move along the lines beside his mouth. He wanted her breath in his ear, her hair in his hands, her flesh against his. He wanted so, he could not think.

"I will see you again," she said.

"No. No, little one. It is wrong."

"I must. I have waited for you all my life." She touched his face and looked into his eyes. Once again her gaze was steady, her face as calm as an unrippled lake. "Give me an hour after they are sleeping and I will meet you behind the wall, there." She nodded in the direction of the bougainvillea.

"I must go back," he faltered. "Up the mountain."

"Then wait for me. I will go with you."

"No. No, you are just a child. You can't. It's not right for you there."

"Then you mustn't leave or I will follow you. Promise me you won't leave."

She was not to be altered. There was a hardness in her determination he had never seen before, although if he could have seen himself clearly, he would have seen that his own hardness was the same. He dropped his forehead to the small hands he still held in his own and nodded. "But we can only talk," he said, and then he could not stop himself from kissing the backs of her hands. He could not resist. Again the excitement flashed through her stomach and down her legs, so that her voice was not clear when she spoke.

"I love you," she whispered. "I have always loved you. I will die for you and I do not even know your name . . ."

Quickly, before it was too late again, he pulled away from her and stood looking at her with pain in his eyes, with sadness.

"Remember," she whispered as he was leaving, "you promised."

"I promised," he said.

She watched him walk away. He did not go across the courtyard to his friends sleeping in the shade, but back out through the gates that opened onto the square. She watched him until he could no longer be seen.

From the doorway of her room, Rita was watching also.

)

Thirteen

For a few moments, Rita was unable to move, or even to think. She stood, blank and numb, staring at the spot where the two had been. Then, slowly, her mind seemed to come to life, and she wondered what was to be done about what she had witnessed. Her heart was in a turmoil, beating against her breast with unnatural force and not regularly. But she knew clearly that action of some sort must be taken.

So she went to the Ave Cantora. He was sitting away from the other men, his back propped against the adobe wall just outside the kitchen door, his cap pulled forward over his eyes and his arms crossed over his chest. She knelt in front of him to see if he was sleeping. He did not look up, but stirred at her presence.

"Hissst," she whispered softly, "García . . ."

Again he stirred. It occurred to Rita that the man must be fatigued beyond imagining to sleep so, sitting on the hard floor against the hard wall in the heat of the afternoon with flies buzzing at his mouth and nose.

"García . . . hiss-st . . . please wake. Please wake now."

He moved his head as if coming back from a long distance.

"Let me speak to you, García. I must talk with you, please." Gently she touched his arm just below the shoulder. Something warm was present in the thin, stringy muscles, something pleasing to her palm. It was as she had imagined it in the church as she had knelt to pray with Father Herrera. A pain of sadness passed through her as she realized again that his warmth was not something she could have.

"Wake up, García . . ."

His eyes opened slowly. His face was blank. His whole body went rigid, and his hands moved to clutch the rifle that lay across his knees. It was a moment, a cloud passing the sun, and then he saw who had spoken to him and his body relaxed. His hands, however, did not leave the black, shiny rifle.

She saw his long thin fingers wrapped around the ugly steel. *Is it warm to your hands as your body would be to mine?* she thought, and the pain grew in her chest, again tightening, as a woman's womb does in childbirth.

She reached down and touched his hand with hers. She let her fingers slide across the ridges the small bones made in the back of his hand, over his knuckles which were scarred, and down around the edges of his fingers until she was also touching the black steel.

"Can you talk to me, please?" she said. "It is very important."

He looked down the veranda at the men, then looked back at her, nodded, and replied, "Give me a minute."

He rose slowly and walked as softly as a cat down the length of the veranda to the dozing men. His boots made no sound on the cement. He knelt beside Pablito, whispered something in the boy's ear and then returned to squat beside her.

"Not here," she said. "I don't want to talk here. It is too important, and we might be overheard."

"They will not listen."

"No, not here. I do not feel free. These things I must say are hard to put into words, and I must be careful in how I say them to you. I want you to understand me. I need your advice, and if you do not understand me you cannot help me with this thing. It is important that I say these things correctly to you, but I am not free here. It is not a place where I can think. You understand?"

He nodded. He glanced back at Pablito, made a motion toward the kitchen door with his head and then rose to follow the woman. She did not touch him, though she wanted to take his hand again, but walked slowly in front of him, turning her head over her shoulder every few steps to see that he was following. She led him past the bougainvillea and out through the small back gate into the alleyway behind the church compound. Then, pausing uncertainly, she asked, "Can you come a little way with me? It isn't far."

She could tell it didn't please him, but he nodded with his somber face, so she led him down the alley, past an intersecting side street of cobbled stones, through a narrow block of pressed dirt, and then between a pink adobe hut and an empty plot filled with stray hibiscus and brambles. They followed a narrow, winding earth track up the hillside into denser vegetation and taller trees, until they came to a small plateau. Ferns and mossy undergrowth covered the ground, and high-standing tempesque trees spread overhead, offering them shade.

"Here," she said. "We can talk here. Is it all right?"

He looked around, nervous as a soldier would be in such a precarious spot, thinking automatically that he was in the open and surrounded by dense growth—not a place to stop and rest in war time. It was the sort of thought that grew habitual to a guerrilla fighter. But these observations passed, and he noticed that the place was beautiful, that birds sang above their heads and small hidden creatures chirped at their feet. He remembered that the world was full of such places, that the earth had other purposes than serving warriors.

He smiled at her. "It's fine," he replied. "I like it. Is it a favorite place of yours?"

Unexpectedly, she found herself blushing. She had come here with Manuel. It was here that they had conceived her son. She was sure of it.

"I come here sometimes," she said, but did not explain why.

He seemed to read her thoughts. "A nook for lovers," he said with a smile.

"I suppose so." She too smiled, then added feebly, "The young ones from the village come here."

He laughed and bent down to pick a wildflower from the grass. "The people who live here are very lucky," he mused, studying the tiny petals. "This mountainside may be one of the loveliest places in the world. I haven't seen much of the world, but I don't think anyone has countryside more beautiful than ours."

"Yes," she nodded, loving the sound of his voice,

but regretting that he had spoken so. Now it would be difficult to tell him what she had come to say.

She sank down on the green turf and waited for him to join her, but he remained standing, with one hand still resting on his rifle. He seemed lost in another world, studying the small flower he held in his hands. She could not bring herself to break into his soft meditation, so she sat silently, waiting for him.

Then he returned from his private thoughts. "I'm sorry," he said, "you needed to talk to me, what was it about?" But he remained standing, cautious.

"I find it difficult to begin," she faltered, "but I know it is important. It is something I can feel. It is not right, but I am unable to begin . . . can't you sit here beside me?"

García shrugged. "All right."

He crouched in front of her, his rifle across his knees, and looked at her with a gentleness that made her speech even harder. It was so different with him. If he were the lover, it would not be wrong.

"I cannot tell you why it distresses me," she said. "I don't know what to do, but I must do something. It is not right, and I'm afraid I cannot even tell you why . . ."

He frowned, "Is it the men? Have they offended you? Have they been rude?"

"No. Of course not. Nothing like that." She twisted her hands in her lap, impatient with herself. "It's nothing to do with me."

"What happened? Just tell me what's happened."

"This afternoon, just a while ago, I came out of my

room. I had just finished my rest, and I came out into the courtyard, and . . ."

"Yes? And what?" He was truly puzzled.

"It was him. He was there."

"Who?"

"It was the one who came before, with Memo, to look at the church. But he didn't come then for the reason he said. He said he came to pray, but he didn't. There was something wrong about him. I felt it at the time. He frightened me. And now he has come back. I don't know what he wants with us . . ."

"What kind of man is this?" He was worried, thinking of soldiers, of spies.

"He is a soldier."

"What is his name? Did he tell you that?"

"No, only that he was a friend of Memo's from the other side of the mountain."

"A guerrilla?"

"Yes, not Guardia, but a fighter. He was dressed like you, something like you . . ."

"When did he come?"

"Two days, maybe three days ago . . ."

"And he said he was from the other side of the mountain?"

"He said nothing, except that he had come to look at the church, to pray in the church, but it was not true. I heard in the village that he looked at everything. He went everywhere. He looked at every house, at every shop, he looked at everybody, even the children, even the fields and the crops. I don't know what he wanted, but I know he didn't come to pray. There is something wrong about this man. I feel it."

"What did he look like?" García asked, his voice guarded now.

"He had black eyes. They frightened me, they were so hard. His eyes never moved, and his face was like that too. It was as hard as a rock. Stiff like a rock. He spoke very softly, but his eyes were unkind. I felt afraid, cold all over, the way you feel when you look at the eyes of a snake, you understand?"

"Perhaps the man was thinking," García replied softly. "What else did he look like? Was he tall, fat, old, young?"

"No one could tell if he was old or young. He was mysterious, I told you. He looks like a piece of stone."

"What else did he do besides look at the church and the village?"

"Nothing. He just looked and then went away."

"Then, Rita, I think he meant you no harm."

"Perhaps not, but I don't trust him. I was glad when he left, but now he has come back again today."

"What did he do today?"

"He was in the courtyard . . ." She faltered again.

"What else? Surely just standing there is not enough to upset you. Do you think he was a spy from the Guardia, dressed like a guerrilla to fool you?"

"No, he came with Memo the first time. I don't think he came as a spy. Maybe he is one of yours even, but I don't like him."

"Did he ask you to wake me? Did he speak to you at all?"

"No. He said nothing to me." She ducked her head, embarrassed and afraid he would laugh at her when she told him the truth about what she had seen. "I don't think he saw me."

García was perplexed now. "Surely there must have been something more than his standing there. You can't have been so upset just by the sight of the man, especially if you think he is one of us. There is more, isn't there?"

She nodded into her lap.

"What then? Tell me what it is. What did this man do that upset you so much?"

She looked at him pleadingly, her face red with shame. "He was kissing her!" she blurted.

At this García cocked his head as if to hear more distinctly and leaned toward her. "Kissing? He was kissing someone?" He had been thinking for some time that the man she was describing must be Beto, but now it sounded as if it could not be him.

"Yes," she nodded, "and she was kissing him, too. That's the terrible part. That's what has made me so worried. I must do something about it. I must stop her. I've told her he is not a man to be in love with, but she won't listen to me. I've begged her and told her stories, but, maybe it is because she is only a girl, and has been locked up there inside the convent walls with no young men to look at or speak to. Maybe it is because of that that she won't listen to me, but I must do something." Rita's voice was growing more intense every moment. "It is wrong for her, I'm sure. And I have to be responsible for her; I'm older, and there is no one else. She has lost her mother, really she never had one, so how can she know anything? She's just a girl. It's not her fault . . ."

"Are you talking about the sister?"

"Yes, yes, Sister Magdalena. She was kissing him, and holding his hand."

García finally sat. He let his long legs sprawl in front of him in the grass. How could this have been Beto? And if it wasn't Beto, who was it? Who else but Beto could have come into the courtyard so stealthily and slipped away without rousing them?

"You don't think he was a friend of hers?" he questioned.

"No. She had never seen him before that day he came with Memo."

"Are you sure?"

"I swear it. None of us had ever seen him."

"And you're sure he said he was from the other side of the mountain?"

"He said nothing. I told you. He only stared like a black rock. Memo said it. Memo said he was a guerrilla fighter from the other side of the mountain, a real guerrilla, from the war. I believed him. The man looks like a soldier, not like you. He is not a poet, like you; he is a soldier. He is a fighter in his soul. That's what makes me so afraid. She is only a child. This is not the kind of man for her."

"Can you be sure they were really kissing, not just whispering so as not to wake us?"

"I am sure." Rita dropped her head and began to cry. It was more difficult than she had imagined, for now she felt not only fear and shame for the girl, but also a feeling she had not expected. She felt envy for the girl's passion, for the fact that it was returned. "It was a vile thing . . .," she sobbed, ". . . he was not kissing her like a sister. He was kissing her like a lover, right there in the open courtyard for anyone to see."

"Did anyone else but you see this?" he asked, wondering even more how it could be true, who the man really was, and why he was kissing a nun.

"No. Everyone was asleep. It was siesta time. It was only an accident that I saw."

"Ahhh," García sighed. He leaned back on his arms and forgot his gun. He stared up into the branches overhead to consider. But he did not express his thoughts aloud. If it were Beto, it was a dangerous thing, but it seemed so unlike Beto, it must undoubtedly be another man. Yet any one of them was in danger if he allowed himself to be snared by a young girl, especially this one. It would do no good for their relations with the village for word to get out that a nun was being seduced. But on the other hand, if a stranger was lurking about, that was also dangerous and must be investigated immediately. He wished that Beto was there to handle the matter, or possibly explain it to him.

He worried through these thoughts while the woman sat silently in front of him, her face moist and drooping.

"If this man is with us," he said, "then we will be able to do something about it. I can understand that you do not want the girl defiled."

She could not let him say this. She could not let him think she believed love between a man and a woman was a bad thing. "No," she answered quickly, "oh, no, she has every right to be loved. It is a thing a girl must have to become a woman, and she is a lonely girl." She hesitated, searching her heart.

"You do not understand, señor. I love her as if she were my own child, and I want only what is best for her, what will bring her happiness. She has no wish at all to be a sister in the church. Her only desire since the time she was a small child has been to marry and have children with her husband. That is all she has ever wanted, and it is all that can bring her happiness. Her happiness is the only thing I want for her too, so I do not object to her finding love with a man. That is not the problem at all. Love between a man and a woman is the most precious and wonderful thing God has put on this earth." Rita spoke with great earnestness now, her words pouring out.

"But, Señor García . . . please understand me . . . I knew I could not say it clearly . . . it is not love. No one can say it is wrong to touch a man, to caress him, to hold him strongly against your body, that is a thing that God has made. He has invented this thing for us to do, and given it all the beauty of the whole earth and all the heavens in one small act, so it could not ever be wrong, and especially for a beautiful young girl with so much hope in her heart, and such a great emptiness to fill . . . No, my friend, understand me, please, it is not the love that is defiling, it is the man . . . only just this man . . ."

He watched her face closely as she spoke, and the depth of her feeling was plain to see. He still wondered if it could be Beto. Beto was the only man he had ever met whose face made one cold, whose face could be like stone, unmoving. Yet it couldn't be Beto, for he would never kiss a girl in public. To Beto a woman was something to be used in the dark places of the night, quickly, then forgotten as one forgets his daily rituals.

At last he said, "I think you had better talk to this girl."

"But she will not listen to me, that is why I've come to you."

"You wish me to talk to her?"

"Yes, please. You are one of them. You are a soldier, but not a soldier. You can say it with words she will understand. You can tell her that it is not love, but the man. You can explain to her the kind of man a woman should love . . . you can tell her she wants another kind of man . . ." She wanted to say, "a man like you," and knew that in another moment, with him so near, she would say it, shame or not.

"Perhaps," he answered. "But first we must know who this man is; we must be sure he is one of ours and not a spy. Do you think she will listen to me?"

"I think anyone would listen to you." He was very close. His chest was so close to hers. She thought she could feel the heat radiate from his thin, hard chest. She thought she could feel it on her breasts, and the nipples tingled in response. She wanted to bare herself for him, her chest, her stomach, her mouths.

García was not aware of her ardor, but he was very curious about the identity of the girl's lover. "I'll talk to her. Don't worry. I'll tell her what you want me to." Though he was looking directly at her he didn't see her, so he was not prepared.

"Thank you, thank you . . ." Because she could not stop herself she touched him. She put her hands on his cheeks, and then they were on his shoulders which were so thin they were like knives under her soft round hands. Her hands moved down across his chest as she had known they would, and the hard straight body felt to her hands as it had felt that morning in prayer, but it was better. She put her lips to his face and found his mouth and kissed him, and kissed his lips one and then the other and then both together and his lips had the forgotten taste she had wanted these many months since Manuel had destroyed himself. And her breasts followed her hands and found his chest. They were large breasts, full as if with milk for a child, thick and heavy, and they felt strongly, they needed as if independent of her other parts, so they pressed and spread out against his chest, until her nipples were like bones against his bones. Hot. Her nipples were hot with their fire, and through the center of her body came the quick telegraph that flashed the bigger fire to the mouth below. And her stomach rolled inside her, rose up, tightened, grew warm and heavy.

"Thank you thank you thank you thank you."

But García was not prepared. He was stunned, for he had thought of her with only the love he felt for all the people of this country. He had felt the love of family, of home, of good things that no longer existed in his life. This is how he had loved her in the morning when she called him the Ave Cantora.

She called him again, ". . . sweet sweet Ave Cantora, sweet sweet sweet . . ."

He did not know what to do. Her hands had found his legs; they rubbed, harder, pushing at his inner thighs.

". . . sweet Ave . . ."

Dimly he knew that she wanted him to make love to her, and he thought of it like a meal to be taken when it was offered, but he thought only dimly, for she had taken his hand from the black metal of the rifle and placed it between the folds of her heavy breasts, and his hands did not need the guidance that his mind did. His hands did not think; they only moved as Rita's hands moved, looking with a life and urge of their own for what had long been absent. His long thin fingers filled themselves with her round breasts. He let them hang in his palms and felt the hard nipples like nails against his skin. He took away her blouse; he parted it over the fine brown breasts and looked. He put his face between them, but the heat there was almost too great to bear. He drew his head back and took one of the nipples in his teeth and caressed the skin around it with his tongue, which was pink on the brown breast.

He was stiff between his legs; his whole body had no thoughts now and asked no questions, but strained toward her, searching for her thighs. He pulled away the cotton skirt that covered her legs and her belly.

". . . ayeiii, que madre, que madre grande . . ."

His hands felt the full flesh of her belly, and then the thighs. They were full too. His mouth felt her. Everything. All of it, making his nose press into the softness until his face was covered by her skin. He drank her. He drank the milk between her legs while she urged him.

". . . ave ave ave oh sweet sweet . . ."

She lay back then, one time satisfied, her blouse around her shoulders and her large breasts filling the afternoon, with her legs spread and wet above her knees. Then she reached for him and drew him to the ground, pulling away his khaki trousers, and she knelt over him and took his member in her plump hands. She rubbed it up and down against her pressing wet thighs and gave it a brush with the black hairs between her legs. She touched it into her wet mouth there, squeezing it and wetting it with her smell while he watched her and grew larger. And then she sank down, bringing her body, every part of it, across the red tip. She pressed him into the cleft of her breasts until he throbbed more, and then she sank lower still and rubbed her face against him, letting the hard one run through her hair, wiping and massaging him with her long thick black hair while her heavy breasts fell onto his thin straight thighs, and the waters came again between her legs.

". . . ave ave te quiero mi ave ave . . ."

She put it in her mouth but it had grown so large she used her hands to love him too. Ah, she loved him while he threw back his head and spread his legs. She rubbed herself against him where she could and against the wet green fern where he was not, until he broke out and swam in her mouth, fluid, hot, sweet song of the Ave Cantora in her mouth with the taste of thick salt and brine running down her cheeks until he was passive no longer, but pushed her back into the leaves and moist grasses and put himself on top of her, wide-legged as she was, and drank the seminal fluid from her cheeks and took it on his tongue from her rough slippery tongue and asked no questions and did not care while his fingers grabbed deep into the pulp of her breasts and his member, larger he thought than the first time, pushed itself into her navel, pushed it up and down against the give of her soft stomach and then lower and lower until the coarse black hairs scratched and teased him, to lower himself yet again and find the squeezing mouth which bit him, wet and strong, into which he shoved, deep, dark, wet, he shoved his tongues, both of them, darker and deeper, while she swallowed him.

". . . ave ave ave mío . . ."

Bones against her bones. Bones to break them.

". . . break me . . ."

To break them into pieces of pain and color and scatter them, lost to themselves, pieces of light and sound in the darkness.

". . . break me . . ."

It happened.

". . . oh ave ave ave ave . . ."

It happened to both of them.

"Ayeeiii . . ."

And everything stopped. All of it died. All of it was gone. Only that light, only that sound.

". . . ave ave . . ."


They stayed together on the green ferns beneath the high trees, filling themselves many times before the darkness came.

)

Fourteen

The church courtyard was alight with torches and thick with men when Rita and García returned from the hill. Rita gasped in awe and shock at the sight of them. The odors of oily smoke, and sweat and garlic and animal flesh roasting knocked her backwards into herself again, and she stood gripping her little stout hands against her chest.

"What has happened?" she cried to García. "Who are these men? What are they doing here?"

Instead of answering, he searched the brown faces shining in the light of the ghostly flares, looking for Beto. At last he discovered his comrade leaning against a post outside the kitchen door. He was smoking a hand-rolled cigarette and looking in García's direction through small slit eyes.

"Go to your room," he told Rita. "I will be there shortly."

"Please, García," she cried, "who are they? What do they want?" But she knew. She knew in her heart why they had come. They had come to destroy all that she knew.

García had lost his smooth voice. It rasped in his throat with the harsh sound of a drunken laborer. "Go to your room, woman. It is nothing to concern you." His worries had returned doubly, and the lush ferns of the mountainside were far away in his mind.

"Tell me!" she pleaded, wanting him to lie to her. "Tell me now!"

"They will not be here long," he said, realizing she was near hysteria. "They only want a refuge, a sanctuary. Is that not what the church is for?"

"Where is Luna?" cried Rita, thinking, *They have raped her. They have killed her; it is my sin that brought it—oh, God, I have sinned.* "Find her," she begged García. "Please find her."

"Go to your room," he commanded, and then, seeing the tears on her face and remembering she was one of the people he loved like home and family and good things all gone, he patted her shoulder and added, "I will find her. She will be safe. I promise it."

"Thank you . . ."

"I will be there shortly. I'll tell you where she is. Now go."

So Rita went slowly across the courtyard to her small room at the back of the enclosure, but in the process her shrewd eyes took in the changes that had occurred in her absence. Boxes of munitions were heaped about like mountains in the dim light of the flares, ominous under their dappled tents of green and beige. The guns, bigger than rifles, as big as cannons to her inexperienced eyes, sat propped against each other in a small row by the door of the sacristy, which stood open to the night. In the light spilling from the doorway, she could see the shadows of moving men. Their voices, low and gruff, were repeating things that made no sense to her, and they were answered by a strange, crackling voice, metallic in sound. She saw the gleam of the wires that twisted like black snakes up the side of the iglesia, the antenna wires for the radio, and not knowing their purpose, supposed they were for some horrible torture.

Rita shuffled her sandaled feet in the dust, making slow progress so that she could listen to the men's voices and catch the tone behind their words. In their sounds she heard both fear and boisterousness, the high-pitched laughter that indicated bravado, nervous excitement, fatigue, tension, anticipation, and fear, again and again, the fear that hid behind the words and the giggles of the young ones. Although she could not understand their words and their machinery, she recognized the emotions that filled their voices.

The panic was rising in her throat like bile, sickening her with fear, and hard behind it was the bitter taste of her own guilt. She had caused this invasion of rebels with her lust. Surely, God, she had. It was her punishment for her sin against the church, against Father Herrera who fed her and her children, and most certainly her dead Manuel who had never had the chance to fight and would be doubly shamed by her choice of lovers. Sweet Mother Mary, had she let this man put a child in her womb on the very spot where she had consecrated her first pure love? She knew then, weak in her knees, weak and dizzy and drunk from loving illicitly, that she had. Her heart broke with fear and guilt. *I've killed us all,* she thought in her shame.

Slipping into the doorway of her room, Rita watched from the shadows. The Ave Cantora was standing outside the kitchen door, sharing wine from a tall bottle with a scarlet faced stranger who was laughing in a lewd manner. Then the Ave Cantora laughed with him, making bitter tears spring from her eyes, for in her guilt she knew they were laughing at her.

*"Slut,"* they are calling me, she said to herself in a tight voice. *"Pig and whore,"* they are saying. *"A hot woman with her brains in her legs,"* they are saying. *Tell him, Ave, go on, tell him, "She'll fuck them all." Tell him that, sweet Ave Cantora!* And suddenly she hated him, the tall lean García, with a vicious intensity that surpassed even her loving. She hated them all, men in battle fatigues with black boots and big black hard metal rifles, and big black penises to shove at her, and into her and past her legs to break her and shatter her even in her soul.

As her hatred grew, all else seemed to slow and stop. The night fell still and quiet around her, and the men's voices receded into the darkness of their corner of the veranda. Behind her, in the windowless, airless room, her children slept in silence, with no sound. Dreamless, they made no mewing in their sleep. It was silence and emptiness all around, so that she could hear without obstruction.

"Yes, I know," she said through her full lips. Her hair grew damp, and she could feel the chill of river water in the night creeping up her spine. "I hear you coming," she said.

In the silence of her numbing guilt she heard for the first time in her life that sound mentioned only in fear-hushed voices since her childhood. In the frozen tableau of the courtyard, the only movement was the low moaning wail which curled and wrapped around her skirts, rising like a mist, cold and deadly on her bare legs. It was the crying in the night of La Llorona, the ghost whose sin is beyond forgiveness, the ghost who comes bearing guilt eternal.

And now, La Llorona had come for her, speaking with the voice of damnation, one doomed sinner to another, begging from one of her own kind the forgiveness she could never have.

"I have sinned," the crying voice in the night crooned. "A poor woman all alone in her despair, I have sinned, and all my children lie drowned by my hand at the bottom of the cold, cold river, where my body lies with them too. Sin. Sin. I, a poor woman in my desperation, have sinned. Sin."

The icy fingers stroked Rita's body, still hot from her own sin, and in this way her guilt spoke to her, and she knew in her heart, which can know things too deep for reckoning, that she, Rita the criada, had crossed the line, like the lonely La Llorona. She knew with her fear that the magical God of the mountains, who can take any form, become any creature, move in any shape, will tolerate many sins when his heart is jovial, but is quick with his anger when caught brooding. He was not like Father Herrera's God of the sweet, empty words, who would forgive anything, the padre said. With the crying in the night of La Llorona, Rita understood that her fate was that of the mountains, not of the church. She knew in her blood that she had stolen one pleasure too many.

Her eyes crossed the courtyard once more and saw again the form of her beloved. Like so much else in life, he had been a trap set for her, and she had fallen into it, dragging everything with her, taking Pasaquina over the edge into nothingness. La Llorona had told her so.

And the first to be destroyed would be Luna, the precious child with the white face of the moon. She would be torn and crushed and scattered like dry leaves by these hard, cruel men. Rita watched them, their sweat glistening in the strange yellow light, their mouths moving like snakes writhing on the wet ground. "I will finish you," she whispered to the men, to Manuel, to the sweet sweet Ave Cantora of her heart. "I will finish you," she said, and it was a solemn promise.

The beautiful Luna with the pale, sweet face of the miraculous Virgin was not, in fact, torn and scattered in the night like dry leaves in a hard wind. Instead she was sitting quietly beside the dry fountain in the square, demurely holding her hands in her lap. She was wearing her habit of black and was as saintly in appearance as she had been on the day of her arrival.

In front of her the stranger paced restlessly, kicking up little dust clouds with the toes of his boots. He stopped abruptly and rubbed his fist against his forehead.

"There is nowhere you can go with me now," he said. "Maybe someday, when the fighting is finished, if we have both survived, there will be a place." He shook his head and looked back at her, an uncharacteristic expression of pleading in his black eyes. "It's all a hope and no more, you know that. Why do you insist? This is not the time for loving and sweethearts and lace bridal gowns. Those are just the dreams of a young girl's heart, but they're not for this place, this time. How can you argue with me when you know it as well as I do? You're only making yourself suffer for nothing."

"Don't you love me?"

He walked away from her quickly, in a marching step as if to leave, then stopped and whirled at the edge of the plaza. "Would it make any difference if I said no?"

"I wouldn't believe you," she replied calmly. "I know you love me. I can see it. It has been true since the first moment you saw my face, just as it was for me. You can't deny that to me. I'm not a fool or a child. I'm no more stupid than you."

"No," he said. "You are not stupid." There was an awful bitterness in his voice.

"And I'm not an ordinary woman."

"No, you aren't."

"In my heart, I am as strong as a man. I will not desert you or be unfaithful or faint when men are shot or speak rosaries at you when they die. And my body is as strong as a man's. I can keep up on any march and climb as well as you—probably better."

Beto shook his head violently. "It is of no matter what you can do. You are perhaps as strong as any man I have, but it does not matter. What is to become of me in the eyes of my men if I take you with us? They cannot respect me if they see I am too weak to separate myself from a woman."

He was shuddering with frustration, but the girl remained calm and unmoved. "Perhaps such a thing would happen," she said dispassionately. "I do not know. I only know that I must go with you." She stood up, as if she were ready to leave at that moment. "I will go."

"I'm not taking you. That's all there is. I'm not taking you." It sounded as if he were angry, but it was only that he was hopeless and unaccustomed to the feeling.

"Then I will go by myself."

"You're a fool!" he almost yelled at her.

"Stop it!" she said, stomping her foot at him. "It is inevitable. It has been willed since the beginning of my life. If you think you can change what is foreordained, then you are the fool!"

"I will stop you," he insisted.

"How will you stop me?" He turned away from her and walked rapidly as if to escape her voice, but she stalked him across the plaza. "Will you kill me?"

"You are a child," he insisted, but as he felt her approaching, he trembled.

"Then prove it," she whispered, "show me all the things I do not know, show me what a child I am . . ."

He spun around to face her. "No, you won't do it. You won't seduce me again. I won't touch you."

But her large black eyes stared at him across the two feet of space remaining between them, stared unblinking in their determination, until he broke. Slowly he reached across the big, empty space that was no longer than his arm, and gently touched her hand. He held it softly in his own and rubbed the delicate fingers while his eyes grew sad and could no longer return her gaze.

"I love you," he said. "You are right, hija, I love you."

"Yes," she nodded, "I know it."

"And you love me," he murmured. "Until today I did not know that love was a real thing. I did not know that it existed by itself, with a life of its own, I did not know that it was an invisible creature which hunts the heart and traps a man against his will." He paused, and an odd look came over his face, as if he were tasting something unexpected.

"I have never loved before," he said simply.

"You could not," she said, smiling softly. "How could you when you hadn't yet found me?"

Then he smiled at her and touched her smooth cheek with his finger. "You are very beautiful."

"I have been waiting for you," she said, "and now I have found you. When you are my lover it will be the most beautiful thing in the world. It will be more beautiful than God, it will even be more beautiful than freedom . . ."

He looked at her still, smooth, lovely face, each line and slope of it, white in the dark night. He had seen so little of her. He had not even seen her ankles. But he knew how she looked. He could see her beneath the black robes, and she was beautiful. All of her parts were perfect. He could see that.

"It will be beautiful," he whispered. "More beautiful than freedom . . ."

"It will be freedom," she smiled, "for me . . . You are my liberator!"

He laughed at the image. "Your liberator! And what will you do for me, when we are lovers? What will you be?"

"Your family. I am your mother and your sister and your daughter." And then she quirked an impish grin at him. "And I shall be your son and your brother and your father and your uncles and grandfathers. I shall be them all for you."

He laughed, and then took her face in both his hands. "Yes, you will be them all. You are my family. You are everything."

"And you will take me?" she asked softly.

He stared at her with a sweet melancholy in his black

eyes and a deep pain in his heart. "Hija," he replied gently, "I must take you, because . . . because, I do not know how to leave you."

He put his arms around her and brought her softly to his chest. Her small face rested against his shoulder, and his chin lay against her head. They stood together like that for a long time. The sounds of the men inside the church courtyard came to them across the night air. A dog howled at the edge of the village and a child cried in its sleep. But no one entered the square to disturb them. They stood like statues, peaceful in the warmth they took from each other.

Finally, he pulled back from her, and looked at her steadily but with great care, as if she were a thing too delicate to bear his gaze.

"Hija," he said almost breathlessly, "just for one moment . . . may I see your hair?"

She looked at him in surprise and then smiled, feeling her own power. "If you will make a bargain with me."

"Yes . . ."

"If you will give me something in return."

He began to grin too, feeling strange inside, as if there were laughter in his stomach. "And what is that?"

"I'll show you my hair," she teased, "for a kiss—oh, just one, I'm not greedy, but just one more."

"Two in one day!" he laughed. "What a wicked woman you are!"

She drew the veil away from her black hair and let it fall to her shoulders. Then she gently took his hands and placed them in her thick, soft hair.

"You may kiss me now," she said.

So he kissed her on her smooth forehead, and then

on her high cheekbones, and then one light kiss above each eye, and finally one for the tip of her nose.

"Have you counted?" he whispered. "You've made a good bargain."

But her lips were smiling and could not answer, so he kissed them at the corners first and then entirely, with a gentleness which was unbecoming to a soldier of the people.

"Ah . . ." she sighed at last, after he had kissed her again on both cheeks and the forehead, "does this mean I'm no longer a virgin?"

He laughed with her and covered her hair again. "In a way," he whispered. "But you mustn't tell."

"So, you have liberated me—can I tell that?"

"It would not be wise," he said more seriously.

He put his arm around her shoulder and they strolled across the plaza toward the church compound.

"Someday soon," she said dreamily, "I shall tell them all. First I will tell Father Herrera. I will describe to him each kiss, demonstrate every sigh, name every caress. What will he do? Do you think he will faint?" She giggled. "I'll make him faint."

He laughed. "Qué mala eres. How wicked you've become."

"No, I was always wicked," she said lightly. "It just didn't show."

As they neared the gates of the courtyard, he drew back into the shadows, and touched her once, lightly. "Until tomorrow," he said.

"I will sleep with you in my dreams," she promised.

"Muchacha," he said as she stepped away from him, "be careful of your mouth."

"I wouldn't tell!"

"No, not you. But the corners of your lips are held very high!"

She put her hands to her face to suppress a laugh, drew her lips into a glum downward curve, straightened her back and looked somberly ahead.

"Very good," he whispered.

She turned her head over her shoulder, her mouth still low and sad, but in her eyes there was a look that revealed everything—her pleasure, her triumph, her desire. Then she walked sedately in through the double gates.

Beto stood outside a long time, staring at the gates through which she had departed. In his mind he went with her through the gates, past the men lounging against the veranda posts, and into her room. In his mind he was her gentle lover and her loving friend, and he held her in his arms, warm and safe against death and heartbreak and loneliness. In his mind he did many things he had never done before. But with his body he stood unmoving at the edge of the square, cold, hungry, unwashed, and alone.

)

Fifteen

Rita was standing in her doorway and saw Sister Magdalena when she entered the courtyard. Instantly, she knew where the girl had been and what she had been doing. Rita was not fooled by the somber face. The girl walked with the gait of a lover, and her eyes glittered with secrets. She walked past Rita with barely a nod, and went into her room, but Rita was close behind her.

"Where have you been?" she demanded, as an irate mother would.

"I've been walking."

"You've been walking! Alone in the village after dark? I don't believe it. Who have you been with? Don't lie to me. I know."

"Then why do you ask?"

For a moment, Rita was unnerved by the girl's cool, faintly insulting tone. Then she turned and closed the heavy oak doors, dropping the bolt into place. "Tell me what you've been doing," she demanded, walking toward Sister Magdalena.

"I've been talking to my lover," said the girl. Her voice was quiet but defiant.

"He has defiled you," Rita said with bitter certainty.

Sister Magdalena held her head quite high and looked back at Rita without blinking. Her promise to keep quiet no longer mattered. "He has loved me, if that's what you mean." Her eyes were hard and bright. "And if you were the one in his arms, you would call it heaven instead of defilement."

Rita gasped as if a blow had struck her. "No, no. Luna, you must not say this!"

But the girl was oblivious to Rita's anguish, and determined to claim, at last, her place as a woman. She would not let Rita make her a child again, to be scolded and then left alone, empty. Not now that her lover, her liberator, had come.

Proudly, defiantly, the girl turned away from Rita and began to disrobe, slowly taking off her habit as Rita watched, stunned.

"Here! Let me see you!" Rita said suddenly. "Let me see what he has done to you." She grabbed for the black robes and pulled them away from the girl.

"Stop it. What are you doing?"

"Has he hurt you?"

Rita pushed the girl backward onto the narrow cot bed, and began tugging at her manta chemise. "Let me look. Are you injured?"

"Stop! Don't you touch me!" screamed the girl as Rita shoved the chemise above her thighs and pulled apart her thin white legs.

"Take your hands away! What are you doing?"

Rita's face grew black with fury, and she grabbed the girl by her shoulders and flipped her over on the bed.

"The blood!" she shouted. "Where is the blood?"

At first Sister Magdalena didn't know what she was

talking about, and only cowered, whimpering, in the corner of the narrow bed. "What are you doing to me?" she cried. "I thought you were my friend. Why do you abuse me?"

"Where is the blood?" demanded Rita.

And suddenly the girl knew what she meant. The blood on the sheets on the wedding night. If she had been loved by a man there should have been blood between her legs. Rita was looking for evidence of her love.

In that moment, Sister Magdalena's mind became sharp and cunning. *If Rita discovers that I am still a virgin,* she thought, *everything will be ruined. They will lock me up and keep me away from him. I must make her believe it is too late.*

"There is no blood," she said coldly. "Do you think this was the first time?"

"Oh, no!" cried Rita, shrinking back from her. "Oh, no! It can't be. Oh, no . . ."

"Yes," said the girl, moving forward across the cot in an animal crouch. "Oh yes, it has happened before, more than once."

"It can't be."

"The first time. The first day I saw him. It was then."

"But that's impossible. You were with me all that day. And he went back to the mountain with Memo. That's not possible."

"He came down again." She paused and smiled mockingly. "You sleep very soundly in the night."

Rita was numb from the girl's words.

"He came back and I let him in through the alleyway. He came here to this room and he stayed with me all night long. Then in the early morning we took the sheets and together we washed them."

"Oh, no . . . oh God, no . . ."

"So you see, I am already lost." The girl lay back on the bed, watching Rita cooly. "There is nothing for you to do."

"Luna, please, you must listen to me." Rita's voice was pleading. "It doesn't matter what you feel now. You are only a girl. You know nothing about love. You have no experience with men. This is not a man you can marry. He is not the kind who can love."

"You're wrong. You're entirely wrong." Sister Magdalena's face was flushed now. "This man is my lover and my sweetheart and my friend. He is the only man in the world that I will ever love. If he dies tonight I will never love another man, but will love him in his grave for the rest of my life. You are wrong. No other man could ever be my lover. No one else could ever be my husband. I have found him, and I will not give him up."

When Rita spoke again, after a long silence, it was as if she were talking to herself. "There is no man you can love, but, oh, God help us, we do, we do love them, and it is a sin. I have felt it, I know what it is to sin with my fingers and my tongue, and my breasts, but you cannot love them. You must not! It will be held against you. It will destroy you. I knew it would. I knew it when I prayed for him. To touch him would bring me destruction and I knew it, oh, God in heaven, why couldn't I leave him alone . . . ?"

Sister Magdalena had covered herself with her robe and risen from the cot.

"Go away Rita. Go to bed. You do not understand. You do not know what you are talking about."

"I know," Rita said, her voice low and hoarse with

emotion. "Since I was thirteen I have lain with men, and I know what they are. God has put them here to destroy us."

"It may be that *you* are destroyed," the girl answered hotly, "but I am not you. I have found what belongs to me." Her voice softened. "It is the dream that vanished when I woke, and now is real; it is my freedom and my life. I will go away with him, and I will live with him and fight at his side, and neither you nor God will stop me!"

Rita reached out to touch her. "I love you, Luna," she said pleadingly. "It is for love that I say these things."

The girl was trembling now with excitement and fatigue and confusion. At the touch of Rita's hand, all her self-control seemed to vanish.

"Leave me alone! Get away from me," she screamed, and she hit Rita hard against the face. "Go! You are not my friend!"

Luna was crying now. She was crying from loneliness and frustration and disappointment, and all the anger she felt at sleeping alone flew at Rita. "You are a vile woman with an evil, unclean mind, and you only torture me because you are jealous. You love him yourself, but he is too young and too handsome and too intelligent for your bed. Go away! Get out! It is not my novio who has broken my heart, but you! You, who have said you were my friend and have betrayed me, and cast shame on my happiness. Get away from me! Leave me alone! Don't speak to me ever again!"

Rita held her stinging cheek and stared at the figure before her, who was no longer Sister Magdalena, nor the sweet pure Luna, but a stranger. They had destroyed her— not her body, for Rita knew the girl was lying, that the man had not been inside her. No. They had taken her soul, her sweetness.

Slowly, Rita turned away and lifted the bolt from the door. She walked through it without looking back, sorrow and hatred so mixed in her chest that there was no knowing which was foremost. She pushed through the men who stood outside on the veranda, and walked, unseeing, across the courtyard. When the Ave Cantora called to her, she wouldn't answer, and though he followed her to the door of her room, she closed it in his face.

)

*Sixteen* )

When at last the image of his love had gone beyond the power of his mind to follow, Beto left the plaza, his rifle slung over his back, his hands shoved into his pockets, and his head dropped low on his chest. From the cantina on the corner, the sounds of a plaintive gypsy guitar and a woman's strident voice calling for her lost lover followed him. He made his way down the narrow cobbled lane that wound upward toward the edge of the village and the foot of the mountain. The dogs yapped and the old men groaned, responding to the restless night with that disturbance one experiences in the calm brooding that precedes a storm.

Beto climbed a short distance up the hillside and found a flat rock. He was just high enough to see the twinkling of the kerosene lamps in the windows of the adobe casa and the faint glow of torchlight that came from the courtyard of the church. He sat down and placed his rifle across his knees, then took a cigarette from his shirt pocket and lit it behind cupped hands.

She had said it was foreordained. She had said it was useless to fight destiny. These were things he wanted to believe, for it made sense of the senseless; it explained his sudden, overwhelming desire for her.

Beto knew there was an element of destiny in life. The war had been his destiny for as long as he could remember. Death, and the moment of its coming, also seemed to be held in the hands of fate. It was something no man could foresee or forestall. Was it possible, he wondered, for love, and a woman, to be in the same category as war and death? Could there be an element of *good* which belonged to him from his birth, as surely as blood, and dirt between the teeth, and wet, rotting feet were his own? He smoked and pondered in the darkness, finding an element of unreality in his life now that the girl had entered it, thinking that happiness could not be as real as horror, and that it must be only a passing illusion—the bright glint of sunlight reflecting off gunmetal that with the shifting of the clouds would vanish as quickly and unexpectedly as it had come.

But even as he knew the happiness was not real, he lived it in his mind. He would be with her tomorrow, just as she wanted. He would be her lover, in secret, in the woods, in the fields, wherever there was a hidden spot, and it would be as breathtakingly beautiful as she had promised. He would set her free from the barrenness of the convent of her childhood, and she would fill him with the family that was dead these many years. If destiny willed it, they would go on together. If not, there would be memories left to fill a space in his heart that had been empty all his life, and he would feel, when his moment of truth arrived, that he had lived fully and known all there was to life.

He tossed his cigarette in the darkness and sat listening to the muted sounds from the village. A small black snake slid across the path. Whether it was deadly or not he couldn't tell in the darkness; all he could distinguish was the shape that undulated in the dust.

*Like the serpent in the garden,* he thought to himself. *This slitherer comes to remind me that Paradise is but a dream. I will not see it. I will fight for it, but I will not live in the time when a man may lie in the arms of his beloved, free of want and fear.*

But even as he thought these things, his heart tightened and rose in his chest. And almost he seemed to hear a soft voice in the darkness, perhaps the voice of the snake. "Why should *you* not have something of beauty?" it seemed to say. "Other men take what pleasure they can find. They take it now, while there is time. They do not give themselves away to a cause that is hopeless."

And he thought, *It is true. The people want their freedom. They want their hectare of land to grow food for their families. They want to be free of the rents they cannot pay, free of the death squads and the army that marches into their worthless villages with tanks and mortars to massacre the women and old men and children. Yes, these things they want, but in their hearts they do not want the fight, the responsibility to make themselves free. They think we are fools to throw our lives away, even though it is for what they themselves want.*

He lit another cigarette and rubbed his forehead with the heel of his hand. *Why do I have to know this? Why not some other? How have I come to this place?* The stillness of the night wrapped itself around him so closely that he could not breathe.

"Beto?"

The whispered sound of his name was like a soft sighing in the silence. For a moment he thought the snake was calling to him, but then he saw that it was Memo. He cursed himself for his carelessness. How could he have let a clumsy boy like this one get close enough to surprise him?

"Beto, what are you doing?" The boy sat down on the ground at his feet. "Why are you not below with the men?"

For a moment Beto could think of nothing to say. His carefully developed ability to speak easily, manipulatively, seemed to have deserted him, and he just stared at the boy. *This was myself once,* he thought at first. But no, he had not been the same as this young one. He had not become part of the revolution for dreams of glory; he had not wanted to swagger with leadership. To be a leader was something that he had come to because his eyes could see where others' could not, because his voice was heard where others' were drowned out. And now, tonight, he was feeling the pain of his burden, a pain that Memo would never imagine.

Memo grew uncomfortable under Beto's staring silence. *What is the matter with this man?* he thought impatiently. *This is not the way of a strong leader, to sit in the darkness as if bewitched.*

At last Beto spoke, but his voice was strange and his words were unexpected.

"Why do you wish to be a fighter in the revolution, Memo?"

*Perhaps this is a test,* Memo thought. But he would have no trouble passing it.

"Revolution is the only way," he began, his voice mechanical. "Only through struggle will we be free. The

188

government and the church would take from the people their life's blood. They rule through fear, and for their greed the people starve and die. We must unite against the arm of the dictatorship. The fruit of our struggle will be victory and a new life for the people of El Salvador."

Beto felt bitter laughter well up into his mouth, but he did not let it make a sound. The boy knew nothing more than the slogans that were taught by rote to village children —fragments of ideas which meant nothing because they were not understood.

"But you, Memo," he said quietly, "what do you want from this revolution? What do you hope to become?"

The boy looked back at him, blinking uncertainly. He had never thought of being asked this question. After a moment, Memo dropped his head and stared intently at the toes of his boots; when he spoke at last, he seemed to be talking to his feet rather than to Beto.

"I wish to be like you, a man who is respected, who is followed even by strong men. If I had such respect, I could do something big, something important. I am sure of it."

"And then you would have even more respect, no? The more great things you did, the more the people would love you, is that not so, Beto?"

The boy nodded eagerly, forgetting for a moment his pose of nonchalance. But he saw that Beto was looking at him sadly—or perhaps not looking at him at all, but looking past him, as if there were a void of nothingness there, staring like a man about to be drawn over the edge, into the void. The hair rose on the back of Memo's neck, and he had to restrain himself from turning to see if a great yawning emptiness had opened up behind him. So when Beto spoke, it was a relief, even though his words made no sense.

"It is not so simple," said the weary voice. "It is not only the government we fight, not only the Guardia, not only the army, not only the rich who rule them all. We fight the people themselves. They see that their own children, for want of food, do not grow, while the dogs of the landowners become fat on meat and milk. And yet, knowing this, the campesinos march day after day to the plantation fields to pick the coffee beans. If you come to them and say 'I will lead you to freedom,' they will not thank you. They will not help you. Only when the death squads are at their door will they cry out to be saved, and they will say 'Lead us out of danger,' and when you have done that, their minds will fall asleep again and they will spit on your heels. And you will have been the one to die; again and again, you will walk into the death of your heart."

When the voice stopped, Memo's mind seemed numb. *I do not understand,* he thought. *These are not the words of a hero. What has happened?*

And in the silence, Beto was also thinking to himself. *Why can't you give it up, fool? Why do you not take the girl and leave them?* But in his heart, he knew that he would go on, bone tired and unflagging, until he died. Whatever his life might have been, it did not matter now. In every part of him now he was a fighter, a leader, and this was his whole being, forever. There was no turning aside.

The scuffling noise of an animal in the brush broke the spell of stillness which had settled over the two men, and for each, it seemed as if his thought snapped like a frayed cord. Without speaking, they rose and started down the mountainside. Memo was still perplexed, but Beto was hardly aware of the boy. He felt strangely at peace with himself now, and ready to begin what must be done.

Seventeen

Rita sat alone for some time in her room. She looked at the mirror, with its gold frame, and thought of the day it had arrived, and the pleasure she had felt. She looked at her own face in the mirror and saw that it was not the same as it had been then, though she could not have said just what had changed.

After a while, she went silently out of her room and along the veranda, ignoring the rebels and keeping to the shadows. She slipped unnoticed into the church and kneeled, as if she were praying, in the very darkest part. From the small room off the altar, sounds could be heard, the odd squawking sounds which came from the radio, and the voices of men talking to each other. She recognized the voice of the stranger, and that of the muchacho, Memo, but not the third voice. The stranger and the third man were laughing.

"Ah, then the radio does not say tonight that the government has surrendered?" the stranger was asking, and the third voice replied, "No, but maybe I have not asked it politely enough, Beto."

Beto. That was the hard-eyed stranger's name. Again laughter, and then, "I will be back soon, Raul, and

take a turn here. I must see about Juan, and the others."
Rita watched as Beto left the church. She heard Memo's
voice speaking to the one called Raul.

"If the radio can hear the army," he asked, "can they
not hear us?"

"They could if we let them," Raul answered. "That
is why we must be very careful. If they heard us, muchacho,
they could find us, and that would be the end."

"But we would fight them," Memo said. "Perhaps
we would win!"

Raul laughed grimly. "We have little to fight with
now, my friend. Not until more ammunition is dropped,
and bigger guns, will we have even a small chance against
them. If they found us now, they would crush us as you
would step on a spider."

There was a pause, and Rita strained her ears hard.
The harsh sounds of the radio continued, but they were
meaningless to her. Then Raul spoke again.

"You hear them, muchacho? They would like to
find us, very much. They look for us, and we must hope that
they do not think to look here. *You* must hope they do
not, for your small village would be only rubble if they
did."

"But why should the army look so hard for only a
few dozen men?" Memo's voice sounded suspicious, dis-
believing, as if he thought Raul were boasting. Rita wondered if the older man would tolerate such a challenge. But
Raul's reply was soft, almost teasing.

"They do not seek a few dozen men. It is one they
want, and they would destroy Pasaquina to destroy him. Do
you not know who they want?"

Rita made herself even smaller in the darkness as she heard new voices at the door of the church. The stranger, Beto, was returning, and with him was a boy no older than Memo.

"Pablito," the stranger was saying, "you must go with our friend Memo here, and he will tell you about his men. We will need them to help us tomorrow. In the morning you and Memo must go to their camp and make them understand what we need." Pablito answered respectfully and left, taking Memo with him. The stranger then spoke to Raul, so softly that Rita could not understand what they were saying. Soon Raul, too, left.

Rita was miserable from keeping so still. Cautiously, she rubbed her eyes and stretched a little, then settled back into waiting. She was not sure what she waited for, but that did not trouble her. All the turmoil which had filled her soul for the past few hours had hardened, solidified, into an inarticulate certainty. The course of destiny was set, and she, Rita, had only to wait and watch as it unfolded. Her own part would become clear.

After some time, Rita saw a dark figure enter the church and move, rustling, toward the lighted door of the small room. It was the girl, her face a pale flash in the candlelight of the altar, her dark hair a flowing mass, for she wore no veil. Rita watched her go into the room to the man Beto, who was her lover even if he had not taken her, and as she watched, she felt nothing except coldness in her stomach. The cold was creeping through her. It had begun when the chill of La Llorona had wrapped itself around her legs, and it was in her stomach now, and already it was reaching into her chest.

"Hija, what are you doing here?" Beto's voice was startled, and yet a sweetness was in it. "You must go back to your room."

"No, I must be with you, camarada," the girl said lightly. "We are to be together, is that not so? I must learn about what you are doing."

"Now is not the time," he said. His words were firm, but Rita could hear that he did not mean them. "I must think, and I cannot do that if you are here."

"What do you think about? Tell me, so that I will know your mind as I know your heart."

There was silence. Then:

"I must listen to the broadcasts of the army and try to think what they are planning. They know we can hear them, so they do not say openly what they mean, but one who knows how to listen can sometimes hear more than he is meant to."

Rita had to think about this very hard, but she concluded that he was saying that although the voices on the radio told lies, sometimes the truth could be discovered anyway. While she was thinking, however, she missed what the girl said next. All she heard was Beto's voice chuckling in response.

"You are more trouble than any girl I have ever known," he said, and he sounded like a boy, not a man with the cold eyes of a snake.

"I am glad of that," Luna was saying, "for now I shall be different from all the others in your mind."

"It seems impossible to make it otherwise," he answered, and to Rita's ear, he did not sound entirely happy that this was so.

"Where is the army now?" the girl asked, and the playfulness was gone from her voice.

"Very near. But they do not know it."

The small silhouette of the girl bent over the radio.

"I have seen radios before," she said, "but not like this one. What do all these switches do, and these lights?"

Rita did not understand what Beto said then. He spoke of "transmitting" and "receiving" and "frequencies," and many other things which had made no sense to her. But at last, Luna said, "Then when this red light is on, it means that others can hear us speak through the radio?"

"Yes," Beto replied, "but they must not. If our transmitter were on for very long, the army could calculate where it is, and they would know where to find us. We cannot risk sending any message now, when they are this near. We must hope they move away so that we can reach our own people and arrange for supplies."

Rita suddenly realized that another figure had entered the church. Someone was standing near the confessional. In the darkness she could only make out the shape, but she knew it, she knew the taut thinness. García was listening, as she was, to Beto's voice. He stood there for some time, and all the while, Rita felt the coldness clutching at her heart, and filling her throat. When at last he turned and vanished through the door into the courtyard, she had become a pillar of ice.

The two murmuring voices in the far corner had grown so soft that Rita could hear nothing further, and after a few minutes she crept to the doorway and out into the moonlit courtyard. As she vanished through the door, the last of the votive candles on the altar flickered out, leaving the church drowned in darkness.

Eighteen

Almost before the sun was properly up the next morning, Father Herrera found himself in the unwelcome company of Dueña Isabel. She had taken the hitherto unheard-of step of coming to see him—indeed demanding to see him. She had sent one of the village boys to announce her impending arrival, and Father Herrera had looked immediately for Rita to handle the matter, but as usual these days she was nowhere to be found. There was no place for him to entertain his visitor, he realized. He could not go into the kitchen where the miserable old man persisted in dying; he could not go into his study, which was now a nunnery; he could not even go into the church, which housed the machinery of these despicable guerrillas. So he was reduced to sitting on a bench in the courtyard while he listened to the piercing voice of Dueña Isabel.

"What is happening, Father?" Her jowls wobbled indignantly. "Who are these men? What are they doing to our village?" The questions seemed to go on endlessly, but that was just as well as far as Father Herrera was concerned, since as long as she kept talking, he was not required to answer her. After a while, however, he perceived that she had stopped talking at last and was waiting for him to reply.

In a burst of good fortune, for which he heartily thanked the Lord Jesus Christ, Rita appeared in the courtyard, and he called out to her at once.

"Rita! Rita! Dueña Isabel is here. Bring us coffee at once." And turning to his inquisitor he said, "You will have coffee, of course, señora."

"This is not a time for refreshments, Father Herrera," she said vehemently, but just as he had hoped, she was spurred on again. "Our very homes are being . . ." her voice droned along as the priest wondered feverishly how he would answer her questions. Indeed, they were not questions, but accusations, and grossly unfair, he thought. How could he be expected to stop these filthy men? They had no respect for the church, and he had no guns to threaten them. He was helpless against this rabble.

Rita appeared again with the coffee, just as Dueña Isabel was reaching a crescendo of complaint. At least the recently-worthless criada had had the wit to bring sweet bread, and so Pasaquina's leading parishioner was temporarily unable to speak because her mouth was full. Father Herrera searched his mind for some topic that would turn her attention away from the guerrilla infestation, but he found himself unexpectedly saved the trouble. Sister Magdalena emerged into the courtyard, and, catching sight of her, Dueña Isabel froze, holding her handful of pan dulce aloft. There followed a moment of blissful silence. The sister walked in a stately way across the courtyard and passed out of their sight before Dueña Isabel could release her breath with a hiss and turn to the priest.

"So that is the 'Virgin' they are all talking about,"

she said avidly, waving her chunk of sweet bread in the direction of the vanished nun. She leaned toward Father Herrera and her voice thickened. "Some are saying that she is a miracle sent by God. But others say she has brought on us the troubles of this war. What do *you* think, Father?"

*This goes from bad to worse,* thought the desperate priest. *Is there no end to my troubles?* He had no more idea what to do about Sister Magdalena than about the rebels, although they were equally pestilential in his view. He would have given a great deal to be rid of the whole lot.

Dueña Isabel was still looking at him expectantly, though by now her mouth was once again full and undulating rapidly. Father Herrera took a deep breath and was about to launch into a suitably unintelligible—he hoped—theological explanation of the whole business, when a tumult of noise preceded los muchachos into the courtyard. Talking to each other loudly, the boys swaggered into the church, leaving both Father Herrera and Dueña Isabel stiff with repulsion in their wake.

Father Herrera was the first to recover, and he seized his chance. "Señora, please, I must talk to these poor boys. They must not be led astray." He stood up and gestured vaguely but energetically, trying to signify urgency. Amazingly, Dueña Isabel rose to her feet, grasped one of his hands and kissed it.

"It is so good to know, Father, that we have you to rely on. You have set my mind to rest this morning that God will protect us."

With that she took herself through the gate and

Father Herrera gave a silent prayer of thanks for his deliverance, although he said to himself privately that it had been long enough in coming. And so, he thought, had Rita, who had only just now returned for the coffee cups and the empty tray. It came to him suddenly that she looked strange, and he wondered if she were pregnant again, but that thought occupied his mind only briefly, crowded out by more imminent problems.

"Rita," he said darkly, his patience exhausted, "if the old man has not died before dinner, then he must do it somewhere else!" Having delivered this ultimatum, he turned and stalked off into his room. As Rita passed him with the tray and cups, he was muttering to himself, "Why can't the army come and rid me of this scum? Where are they when you need them?"

Sister Magdalena was impatient. She had found her novio at last, but here she was, still in the clothes of a nun —stupid clothes that were too heavy, too warm to drag around, and exceedingly plain—with nothing to do but wait. He was nowhere to be seen this morning; in fact, almost all the men had vanished. As she did not want to encounter either Rita or Father Herrera, she had determined to walk out of the village and into the trees on the hillside in search of Beto. It was an exciting idea, for she had never done such a thing. But this was the beginning of a new life.

She walked through the village, ignoring the shocked stares that followed her, and began to climb along the first pathway she saw. Her skirts got in the way, and after only a few minutes she was frustrated and a little tired, but the sensation of walking alone out of doors was so novel that she did not mind her discomfort.

It was not very long before she realized that she was not alone on the path. Coming along behind her, and rapidly drawing closer, was the man called García. His thin legs were so long that he seemed to cover the distance between them almost effortlessly.

"Good day, señorita," he said, and the sound of his voice was unexpectedly pleasant. "May I walk with you?"

García did not seem at all surprised to find a nun climbing the path above the village. In fact, she had the impression that he had expected her to be there, had perhaps even followed her.

She answered him primly. "If you wish to know the truth, señor, I had hoped to walk alone. I have not had many opportunities to do so."

"Perhaps that is why you do not realize the dangers. It is not safe for a woman alone, not even a nun. Especially not a nun, for it is so odd a thing that some might think your habit was a disguise."

"But it is a disguise," she said without thinking.

"What are you then?" García asked. She did not hesitate in her answer.

"A woman." As she said the words, she could feel the lips of her beloved on hers, the touch of his hand, hot against her own.

Although she did not realize it, García could read her thoughts in her eyes. "Please," he said, "will you not sit with me a moment?"

With a mixture of curiosity and reluctance, she nod-ded, and sat down on a large stone at the edge of the path, draping her skirts around her. García sat on another stone a few feet away.

"I too am in disguise," he offered. "I look like a soldier, but that is not the whole truth. Perhaps you can guess what I really am."

"I do not need to guess," she said quietly. "I have seen others like you, in my father's house when I was very small." Her eyes rested on his. "You are a man who dreams."

*So this is the one who has changed Beto,* García was thinking. *Seeing her now, it is not so hard to understand after all.* Her face was so beautiful no one could be unmoved by it. And there was more—a strength of will that matched even Beto's, and a strange knowingness. *But still,* he thought, *it is bad.*

"I dream, yes," he answered slowly. "I dream sounds into songs. And I dream that one day my songs will be the poetry of a free, proud people. But that day will not be soon, I tell you. We will none of us have our dreams in these days to come."

"But if we renounce our dreams," she protested, "then what is the use of fighting at all? We might just as well remain the tame animals of the coffee barons."

García shook his head gravely. "It is not renuncia-tion I speak of. Although we cannot live our dreams, we can hold them in our hearts, to grow dearer and more true."

"And if we die with our hearts full of unlived dreams? What then will our lives have been but shadows?"

"Our lives will have been monuments, señorita, not

shadows," García answered. "We will have proved that a few, at least, will sacrifice their own deep desires for the sake of a people. And our dreams will live far longer than if we had squandered them in the pleasures of a moment."

The hillside was alive with small sounds—the hum of insect wings, the click of cicadas in the grass, muted rustlings in the branches of trees. Yet Sister Magdalena heard none of this. She was intent on her own thoughts, and her concentration created a bubble of utter silence around the stone on which she sat. García could almost see the stillness that enveloped her. He too was still, waiting. He watched the beautiful young face and saw there fleeting signs of struggle, of confusion, and finally of resolve.

"It is no use, señor," the girl said at last, her voice strong and level. "You are wrong. Our dreams *are* our lives. Without them our hearts are dead, and the dead cannot fight a war. It is not by freeing others, but by freeing myself that I become a revolutionary."

As she said these words, a strange thing happened. In that moment someone who had been living inside her burst into being and was speaking through her lips. "Love is the great liberator, señor," she continued, "and I have found my way to freedom. Do not try to convince me otherwise."

García looked at her a very long time before he answered. "I will try to convince you of nothing. Remember only that he who buys freedom for another often does so at the price of his own life. Think, señorita, then think again, of what you are doing."

Gathering her skirts, she rose and walked away from him, back toward the village, her head held high, her dark tears hidden. *Perhaps we are both poets in disguise,* he thought as he watched her descend. *But still, it is bad. It is very bad.*

Nineteen

The afternoon had worn away by the time Beto returned
to the church. He had begun the hard work of persuad-
ing, or at least subduing, the villagers; for good or ill,
Pasaquina would join the revolution. It was all just as
usual. Most of the villagers were vaguely on the side of
the revolution, but they would not take it seriously until
they found themselves, their homes, their small, poor bit
of land threatened by the army. When that happened,
some of them would rise up in anger and become brave.
But in the meantime, they would all behave according to
their natures, whether foolish or kind or lazy. Three or
four, perhaps half a dozen, with a little money and hopes
of getting more under the old system would be actively
against the guerrillas. They would talk to each other about
the evil invaders, but the people would ignore them, and
it would not matter in the end. There were few here who
could offer any resistance, and none likely to take the
trouble to betray them.

Still, it was not possible to be sure. Beto knew that
he and his men had to control the town as tightly as possible
until they had achieved their objectives. He was troubled
by the nearness of the army. They were not yet in the valley,

but they were closer than he had expected them to be. By now they should have been drawn off by the men he had sent to the south—a far larger group than those few he had with him here. He had counted on the army's being too occupied by this diversion to suspect that he himself had slipped away to prepare a new base. But if the army remained in its present position much longer, it surely meant that they were not randomly searching but *knew* he had come this way. If that were true, something had gone terribly wrong.

Still, there was nothing he could do for the time being, so he forced his mind to concentrate on other matters. The priest, the foolish little Herrera, he had ignored completely, and that was of course a mistake. In many places the priests were friends of the revolution, for their own reasons. Father Herrera was not one of these, Beto was sure, but even so, there might be something to gain from him. Beto was about to seek out the good father for a talk when Raul came toward him across the courtyard. The man's scarred face was sad.

"So Juan is gone," Beto said when his comrade was beside him. Raul nodded. *How many times have we two had this moment?* Beto wondered to himself. Dozens surely; hundreds, perhaps. Who could remember? "Well, it is done then. There is no helping him now."

"No," Raul agreed. "For Juan the pain is over. But for the others it is made worse."

"They have seen men die before, Raul. This is nothing different."

"You are wrong, Beto. You are not thinking. When someone is lost this way—with no glory, a stupid accident—

it always hurts more. And they have time now to feel their hurt."

"Then we must keep them very busy, my friend. There is much to be done before Pasaquina is secure."

Raul looked even more worried. "That is another thing. There is something about this place. We all feel it. Something bad. We are not wanted."

"We are seldom wanted," Beto answered grimly. "This also is nothing different."

But Raul insisted. "No, Beto, it is not the same. There is a bad feeling about this place. It is not just that the people are not yet with us; there is something else."

Beto pondered this. His comrade, usually a calm, brave man, was clearly agitated. Looking around the courtyard, Beto saw a few of the men, staring tiredly at their feet or holding their heads in their hands. They were demoralized.

For Beto this all seemed strange and sudden. He had been troubled since they had come to this place, it was true, but his trouble had also brought a pleasure more intense than anything he had ever known. He had thought little of his men in the last hours, so intent had he been on this strange beauty which had befallen him.

"I will speak to them," he told Raul, but Raul was not satisfied. He had been watching Beto's face and had seen something there which troubled him even more than the death of Juan or the fated miasma that hung over Pasaquina.

"Beto," he said urgently, "perhaps we should leave this village. It is no good for us to be here. I believe we have followed the wrong course and must turn back before it is too late."

"You talk like an old woman," Beto replied, his voice tense. "We have not stumbled in here by chance. We are here for good cause. No, Raul, we will stay until we have carried out our plans."

Only a moment was gone before Raul replied, but in that time much passed in the mind of each man. In the air between them there came a gathering of energy that seemed about to form itself into a challenge, but then, in the last instant, it dissipated, like a fire smothered in the rain.

"There must be a burial," Raul said quietly, and Beto's face relaxed.

"Yes, tonight." Beto patted the other man on the shoulder. "Come, my friend, let us go and see him."

Inside the kitchen, the boys had arranged candles around the old man's body. Juan lay on the cot with his hands folded across his chest. He still wore the bandages, and his face had been cleaned but not shaved. A sprig of bougainvillea lay beneath his hands, and the vivid red against his pale dead skin seemed to screech like a puma.

"Why give him flowers?" snapped Beto. Suddenly he hurt for the old man, who would never return to his wife and grandchildren. All the tension of the day broke over his heart like a summer storm.

"Why candles, hombres? This man is dead! He can't see them. He'll never see anything again!"

"Not now, Beto," whispered Raul. "They are frightened."

"This is superstition." Beto paced around the old man's withered body. "Next they'll be wanting last rites."

Raul shook his head with a wry grin. "Amigo, they did that this morning . . ."

"It is not this place which drags the men down," Beto said angrily. "It is their own weakness, to believe that the dead are helped by candles and flowers and priests. The dead are only what they have done in life and that is what must be honored." Suddenly he swept the sputtering candles to the floor.

"You!" he said to the men standing near Juan's body. "Take him outside. Put him on a stack of ammunition. Put him high. Put him as high as the church steeple, so everyone can see him. I want them to see him. Now!"

So Juan's body was carried out the door and through the hushed men to the center of the courtyard, and placed silently atop a tall stack of munition boxes. Beto yanked a flare from its mounting on the veranda post and strode out behind them. He leapt atop another stack of wooden boxes draped in drab green netting and stood, spread-legged, defiant, holding his torch high. The oily flames guttered in the evening breeze.

"Listen to me," Beto cried. "Listen to me all of you! You will not fall to your knees. You will not bow your heads! Stand up! Be tall! Be proud! This is my comrade, Juan. This is Juan Regalado, a brave man, a fearless fighter, a soldier of the people! Open your eyes and look at him! Open your ears! Listen to me now!

"I am speaking to you about my friend Juan who has died this afternoon, who has drowned in his own blood today. This is my friend, who has stood beside me, and fought beside me, and marched with me, and drunk with me, and laughed with me. And he is dead!" He waved his torch over the empty body for all to see.

"He is dead this afternoon. His head is broken and his face is torn. His lungs are ripped into shreds by his ribs.

"Do you hear me? This man who is my friend is dead! It is the end of him. He is gone. Finished. That is all there is!

"Hear me! He is not with god and the angels. There is no god. There are no angels. There is only—this!" and he pointed the torch again at the small, dead, empty body. "And when this is broken, the man is gone. He will never return. There is no more."

He shifted his weight on the boxes and glared at them, while they stood, frozen, their faces skull-bare in the golden lights.

"We have only the moments in which we breathe. And then we are finished, nothing, it might never have happened.

"But this, my friends, was a brave man, and while he breathed, he took it all, everything there was to take. This man was a lion of strength, a river of courage; he was straight in his back and strong in his legs. He begat many children. He lived a long life, and every breath he took was deep and every sigh was loud and long."

And they answered him in their hearts. *Beto, sweet our Beto, cry for us, our Beto. Weep and die in your heart, our Beto, that we can follow.*

"Stand up tall! Stand up and be proud! This fine man was your friend and while he breathed he honored you with his presence, and now, when he breathes no more, he honors you with his dying."

*Yes, yes, our Beto, speak to us. Cry. Be cut in your heart that we may hear.*

"In his death he honors you, he respects you enough to die in your midst, to remain your comrade even in his most private victory."

*Yes, our Beto, yes, be cut in your heart with our pain, that in your blood we may rise. Give us the words, our Beto, sweet, our Beto of the eyes that see.*

And in his heart Beto felt their pull. He knew their need. So, standing there, quivering in the grief he abhorred, his heart, though he fought it still after a lifetime of defeat, answered to them and said through his deep black eyes, *Have I any other? Is there another path, my friends, my children?*

He felt them in his body and their longing was powerful. *There is none,* he knew. *I have been given no other.* And so he continued, and they drank from his soul in the deepening night.

"Juan Regalado lets you witness his liberation! His liberation which has come at last!"

*Aaahhh, talk to us, Beto. Take us out of our nameless, faceless numbers that lie to no account in the ditches, take us with your vision beyond the corpses too numerous to count, the corpses with faces gone to acid, the dead too thick on the land to remember. Oh, Beto, cry for us. Cry and give us the death that sings.*

So he cried. He screamed aloud until his throat was parched with the noise of it.

"Stand up and salute this man for he has taken the one step beyond. Now he is free—he has finished his fight with honor—Salute him, and—look at me!—vow in your hearts on this brave man's body that you will not dishonor his memory, that you will fight and die as well as Juan

Regalado—with honor, with courage, with dignity—and that while you breathe you will honor his memory with pride. And in that pride of his memory, you honor yourselves as well."

Beto turned toward the sleeping body that lay on top of the munitions boxes, and held his torch high above his head.

"Mi amigo," he cried, "I love you. I love you with honor. Mi amigo. I salute you in your death!" And he released the screaming bullets of his black rifle into the black sky.

In the courtyard an animal yell rose up: "Bravo! Bravo! Bravo, Juan, te adoro! Mi amigo, I love you" while the black wicked bullets ripped the night to tatters, while the small pellets of death raced into the heavens to salute the dying of Juan Regalado.

"I love you! I salute you, Juan! Te adoro!"

And they howled and danced around his body and shot their guns and passed the long brown bottles of chicha until they were drunk and puking and weak with weeping for joy at the death of a brave man, a man of honor.

On the other side of the courtyard, in the room which had once been a chapel for some devout, long-dead priest, Sister Magdalena crouched on her knees beneath the bed. Her eyes were wide with terror, her face whiter than possible; her small hands gripped each other over her chest, while her lovely lips hung loose and wet and red. She muttered incoherently to herself, "Ayeiii, no, Papa, the sounds, no, no, make them stop, make them stop, ayeiii no, mi papa, please Papa, they hurt my ears, make

them stop, ah mi madre, Madre, make them cease . . ."
And all the while the mad scene of bullets cutting through
ripe flesh swam before her eyes, and over and over again
her mother said the prayer which would never finish, the
prayer which stopped short and began again, but would
never end.

Rita had been watching quietly from the door of
her room as Beto shouted in the torchlight and his men
swayed to the rise and fall of his passionate words. She
was not interested in what he said, but only in watching
him. His cold, stone face was changed into something terri-
fying, but Rita was not afraid. There was no fear left in
her.

At last the men lifted Juan's useless body and took
it off into the night, into the mountains. Utter stillness
descended over the courtyard. Rita stood, breathing in the
night air, and soon the air, heavy with the smell of gunfire,
began to change. At first it was sweet and fresh, but that was
only for a moment. Soon the smell of evil and decay en-
shrouded her.

"Susto," Rita whispered. Susto, the bad air that
brought sickness and death. She stood still, as the bad air
flowed over her and into her, and into the village of
Pasaquina. Then she heard his cooing.

She had expected this. La Lechuza sat, blinking, sil-
very gray in the torchlight. Blinking and staring, cooing
softly, La Lechuza, the owl, had come to deliver the message of doom. *Where the owl alights, death comes.*

Rita felt a sweet satisfaction as she looked at La

Lechuza. It would not be long now. The winding paths of life are inevitable, and she, Rita, found herself at the center of what must happen. Her moment had come. All around her, destiny wrapped itself through the village, entwining the houses, the people, the animals, and waiting until she, Rita, made the tiny move which would set the rest in motion. She stood a moment longer, feeling the thickness of the inevitable as it pressed against her skin and took the air from her lungs. Then she walked slowly across the courtyard, her soft body cutting through the heavy doom as through a dank mist, and the doom closed behind her as she passed.

She walked to the door of Father Herrera's room and knocked firmly.

From behind the thick slab of oak, the priest growled, "Go away!" for he was angry and brooding, sick of charity and Christian compassion, which could not be expected to extend this far. The mad rabble in his courtyard had staged nothing less than a pagan orgy, and he could do nothing but cower.

"It is Rita, Father. I must confess."

"Stupid woman. Go away!"

"You don't understand, Father. Please, you must hear me."

He growled and scuffed his sandals against the dusty flagstones as he crossed to let her in, though it was not from compassion but from curiosity that he admitted her.

She fell to her knees and kissed his hand. "Father, I must confess. Tonight."

"Don't be a fool, woman. You can't confess here,

and you certainly can't expect me to hear you in the church. It's been taken over by unwashed mobsters and rabble rousers."

"No, they have all gone. The church is empty."

"They will come back," he said crossly.

"Not for a long time. They have gone to the mountain to bury the old man. Please, you must hear my confession now."

The priest was in no mood to be comforting, but there was always the chance that the criada would say something useful to him—or that her sin would be serious enough to allow him to vent his frustration by heaping great penances on her foolish head. So, still grumbling, he trudged off to the church with Rita silently behind him.

When they entered the church, it was dark and empty. *This is as it should be,* Father Herrera was thinking. *My church should not be violated by these vermin.* But scratchy, faraway sounds could still be heard from the radio.

Rita went before the priest into the confessional booth and leaned her face against the wooden grillwork. He in his turn put his ear to the latticed wood, thinking, *They are nothing but rats in a corn crib.*

"Father forgive me, for I have sinned . . ."

*Next they will have the butter and the cattle and the pigs. Soon there will be nothing left.*

"I have been with one of the guerrillas . . ."

*When these fools are finished with our campesinos, corrupting them, there will be nothing left.*

"He was a lover such as every woman dreams, Father . . ."

*Every one must stay in his proper place if there is to be order. These wild men would turn everything upside-down.*

"And I carry his child; I know it is in my body."

"What?"

Father Herrera suddenly heard what the criada was saying.

"What?" he repeated, not keeping his voice down at all. "Whose child? What are you saying?"

Rita's voice continued serenely. "I am not the only one who has been seduced, Father. The nun, too, Sister Magdalena. I have seen her kissing with the leader of these men."

"How can this be?" exclaimed the priest. "The guerrillas are kissing the nun?"

"Only the leader, Beto. He is the one the army searches for. This Beto is an important man, wanted by the government for many crimes."

"And you have seen these two together?"

"Yes, Father. They have been together everywhere, even here in the church. I listened as they spoke words of love. But there was more." She paused seductively, her eyes glittering with inspiration. "He told her things about the radio. There is a red light. When it is on the radio can talk to others, to the army, Father."

There was a silence. Rita could hear Father Herrera breathing hard, panting. After a little time, she spoke again, her voice sweet.

"Please, Father, what is my penance?"

"What? What?"

"My penance."

He could barely restrain his impatience. What had her sin been? He could not remember now.

"Whatever I gave you last time—twice that. Now go and let me think!"

Rita rose quietly and left the small enclosure. She walked into the night, seeing nothing, thinking nothing at all.

Twenty

When the men returned from their grim revelries, they fell ingloriously into heaps and slept like the dead man they had laid to rest on the mountain. Their rifles slid to the ground; they moaned in their sleep from grief and exhaustion and fear. They slept until the sun was high in the sky and the frightened villagers had been awake and watching from behind their shutters for hours.

García was the first among the men to rise, for he had drunk lightly and remained more disciplined than the rest. His emotions had not needed purging so desperately as theirs, perhaps, because of his encounter with Rita. Stretching, he grinned amiably at the forms of his comrades. He noticed the sun was high and the day well gone, and noticed also that neither Rita nor the priest nor the young nun had ventured into the courtyard. He thought it strange, but not overly important, and shrugged it off.

García put on the water for coffee and found cold beans in a covered pot and tortillas behind the curtain in the corner of the kitchen. The morning was eerily quiet, but García was not a native of Pasaquina and so he failed to notice that even the dogs were silent. He could not know that the morning sounds of the village should have penetrated the thick adobe walls of the courtyard, making a soothing music of cart wheels and horses' hooves and recalcitrant children and bickering wives. So he was not disturbed, and quietly made his coffee in the tin pot and warmed the beans. He dipped the tortilla into the bubbling beans, wrapped them securely inside, and took a lump of dark brown sugar for his coffee. Then he walked out into the sunshine with his breakfast. While he ate, he glanced at Rita's closed door and remembered her smooth flesh and firm hands. It brought a smile of pleasure and appreciation to his lean face, and he fantasized visiting her again many times.

*It is as I thought,* he sighed to himself. *The peons are the very best people in the world. The women are as rich as cream and as sweet as honey. There is no educated woman on earth who could love like that.* And he laughed gently, thinking how lucky he had been to attract her. *It's a shame we'll be here so short a time . . .* But he consoled himself with thoughts of the afternoons ahead.

He threw the coffee grounds left in the bottom of his cup to the dirt and crossed to the well to draw himself a dipper of water. It was cool and sweet and the morning was not yet overly hot. He grew philosophical, thinking of Rita. *If you have nothing, and no hope of getting anything, you are content,* he thought, *but if you have something . . . it is never enough.* Then he laughed at himself.

Two of the younger boys were awake now, and they joined García at the well.

"It is a fine day, is it not?" he said to them, so jovially that they were startled. Neither boy felt at all well after the night's excesses, and García's good cheer seemed somewhat unnatural. But they tried to look alert, so that the older man would not notice their condition. It seemed to work, for García continued his monologue, oblivious to the state of his bleary-eyed companions.

"Never underestimate the effect of a woman, boys," he told them with a wink. "A woman can make miracles for a man if she wishes to."

The boys grinned and winked in reply, although only one of them could guess at García's meaning, having been with a whore—a birthday present when he had turned fourteen.

The sun had reached its zenith by now, but the morning remained preternaturally still. After a few minutes of banter with his two young comrades, García headed for the church. Something strange about the morning had begun to penetrate his mind, though he could not say what it was. But he was curious about what was happening in the world beyond Pasaquina, and his long legs carried him toward the sanctuary and the radio. On the way he stopped to speak to Raul. "Hey, hombre, did you sleep well?" García asked.

Raul rubbed his eyes and grinned. "I drank well, amigo, that's for sure."

García grinned back at him. "It'll give you strength. Come, let us see what the radio has to say this morning."

Raul entered the small room first. He was still half asleep and suffering from a headache, so at first he did not believe what his eyes saw. He blinked and looked again. Then with a loud cry, he slammed his fist down on the table.

"Sweet Mother of God! Look at this! Holy Jesus, what have they done to us?"

"What is it? What are you yelling for, hombre?" demanded García from the doorway behind him.

"Look at this! Oh Sweet Jesus, they've killed us!"

"What, man—what is it?"

"The radio. The goddamned radio. The fucking radio. Can't you see it, man? The light is red."

He grabbed the microphone from the table, and then threw it crashing to the floor. One glance had told him the story. "The damned button has been jammed. The thing's been transmitting all night. Mother of God, they've found us now. Oh Sweet Mother of Christ, we're dead men."

"What are you saying, man! Speak to me! Who has done this?"

"How the hell do I know! The goddamned thing's sending signals around the world, that's all I know. They could find us in Guatemala, in Nicaragua, hell, they could hear us in Cuba—what more do you want, man, we're finished!"

"Has somebody been transmitting on this thing?" demanded García.

"I don't know. I can't possibly know that, but I'll tell you this, hombre, somebody sure as hell has been trying."

The two men were silent, stunned. They stood staring at the radio, its red light like an eye gazing back at them with mocking serenity.

"We must wake Beto," García said at last.

"Good God in heaven," said Raul, "what we must do is get the hell out of this fucking town."

But García was already vanishing through the door and in less than a minute he had returned with Beto.

"Tell me what has happened here," Beto said, his voice low, his face dark and hard, as it had been all his life until he had kissed Sister Magdalena.

Raul had recovered himself somewhat, and he spoke quickly and efficiently. Beto seemed to listen without feeling, but behind the stone face, his mind was racing. Who in this village would know how to work a field radio? Could one of them have figured it out? The priest was probably the most intelligent of the lot. But why would the priest want his village destroyed? The fuddled little man had hardly left his room since they had come. Would he have had the nerve and the brain to do this thing?

Maybe there was a spy in the village, someone who knew more than the others and sympathized with the government—perhaps even worked for the Guardia. It was possible, even in this remote place. Possible, but not likely. And there was another explanation, so simple, so obvious.

"Beto," García was saying urgently, "Beto, we must leave. We must warn the villagers, and then leave. We can do no more. We are too few, with too little weaponry. We must go!"

But Beto stood, unmoving, unhearing. García saw that his friend was in pain. He moved to Beto's side and spoke in a low voice.

"It is not your fault, amigo. We cannot be watching every moment, with no time for grief, for love, for madness. It is too much. Do not blame yourself that for a few hours we were human."

But in Beto's mind, a blackness was growing, a strange, sharp blackness that seemed to pierce and cut his every thought. Words formed in his head and each was like a fragment of dark glass, jagged, slicing through his brain. *I have always known it, from my youth. A man is killed by love. Why, mother, did I forget now?*

He turned stiffly and left the small room. As he crossed the dim church, García was behind him, imploring.

"Beto, there is nothing more to be done here. It is finished. We must go now, while we can still save something."

Beto did not answer, but only walked across the courtyard, his face the face of a man who has seen himself swallowed by the abyss. Around the courtyard, the men stopped and became as statues, paralyzed by the sight of this man who had been their leader but seemed now like an apparition out of hell. Picking up his gun from the place he had left it, Beto walked on.

García turned desperately to Raul and whispered, "Tell each man to take what he can. They must go in small groups, in different directions. As they leave the village, they must warn the people, but we can wait for no one. Go quickly. And if any survive, we will meet again in seven days, where Juan fell."

Raul nodded, and for a moment he held García's eyes with his own. Then he hurried to the men, gathering them with gestures and drawing them into a tight knot near the gate.

García moved swiftly to where Beto was standing, and then García was beside Beto, in front of Sister Magdalena's door. The thick door was bolted inside, and from beyond it, nothing could be heard. Beto lifted his rifle and brought the stock against the door over and over, pounding loudly, but still there was no sound within. He stood back from the door and aimed the rifle at the point where the heavy beam held the door, and fired again and again. When at last he stopped, a banshee's cry came from behind the doors and rose into the warm, gentle day.

Beto kicked the door and the splintered beam gave way. At first no one could be seen inside the shadowy room, and García imagined for a moment that the girl herself, dissolving into her own desperate scream, had floated away into the sky. But the sound of muffled sobs and incoherent babbling came from the corner of the room, and there Beto found the girl and dragged her from beneath the ragged cot. Her eyes were vacant and her long, silky black hair was wet with tears and streaked with dirt from the floor.

"You have betrayed me." For a moment Beto's voice was controlled by the discipline which had been his life. But then it broke. "You have killed me, whore!" He slapped her, his rough brown palm hard against the smooth white slopes he had longed to caress. "Why have you betrayed me, slut? Why?"

The girl's face was uncomprehending. "No no no no . ." she cried. "Oh please no make them stop the noises, make them stop shooting, make them stop the bullets, oh they are hurting me, they are killing me, the noises, oh the noises, the sounds . . ."

"You have brought death on us all," Beto shouted at her. "Whore of the devil, you have destroyed us." He hit her again, but her small voice went on, murmuring brokenly, like water passing over rocks in its rush down the mountainside.

"No no no, ah mi papa, ayeiii, no mi papa, stop them, stop them make them stop oh Father. I cannot stand it, I hate it, oh the noise of it, please, how it hurts me, the sound, please Papa . . ."

García could see that the girl was incapable of understanding what was happening. And Beto, he realized, had lost the mind of a soldier and become only an animal in pain, a lover betrayed. García placed his hands on the shoulders of his comrade and spoke softly in Beto's ear.

"Amigo, please. This thing is wrong. See what you are doing. She does not understand."

But Beto shook him off violently. "She will understand, I swear it. She will know what she has done." He grasped her small arm and pulled the whimpering girl across the rough floor, through the doorway and into the courtyard. Her fingers clutched the edge of the bedsheet and it trailed behind her, a poor unblooded tatter of white against the stone.

Beto dragged her into the church. He flung her against the table which held the radio, screaming, "This is your doing, bitch. Look at it, look at the death of many good men." His face was black with rage and pain. "Fucking, fucking whore!"

The girl raised her head. She saw the radio, but in her mindlessness she did not know what it was. Yet the bullets had stopped. The sounds in her head were diminishing. She turned, and saw through the door the bank of votive candles which flickered sadly, with no magic, beneath the plaster Christ. And then she saw his face, the face of her lover, of her novio, the face she had waited for all her life, the face she had dreamed of since she had been a girl at her mother's knee. The beautiful, incomparable face of her heart, chiseled from the volcanic rock of her homeland.

"Mi novio, querida mía," she sighed, "I have done nothing. I have done nothing, mi corazón."

"You have killed us! You have sentenced us all to death this day." He raised his hand as if to strike her again.

"Ayeiii, stop it. Oh stop it please, querida mía, I have not. No stop it mía, I have done nothing . . ."

Again García tried to make Beto hear him. "Beto, it might have been anyone. I think she has not done this thing. Look at her, man!"

Beto did look at her, at the soft, white face like the full moon, the sweet face he had kissed to set her free. *It is fate,* he heard again in his mind. *It has been foreordained.*

"It does not matter whether she has touched the radio," he said heavily, turning to face García at last. "She has destroyed me, and that has doomed us all."

García took Beto's arm and pulled him through the door, into the dimness of the sanctuary. But the girl followed them. She fell in front of Beto and embraced his knees, her gentle face against the rough green cotton of his trousers.

"No, oh no, my lover, my darling, no I have not. Do not speak to me like this. I have loved you. I have only loved you." She turned her face up to look at him, her eyes pleading. "Mi novio, love me, oh love me again, for I have done nothing. I could do nothing to harm you, for I have loved only you. I will die for you. Ask it and I will die. I love you, believe me, I have only loved you."

He kicked her away, the black steel-toed boot embedding itself in her soft flesh, and García was engulfed by a wave of sickness and despair. This was worse than all the deaths they had seen, all the enemies they had fought. Beto's soul was dying and it screamed out in its agony.

"Ayeiii," said a voice in García's ear, "he has gone mad." Turning, García saw the boy Memo, his face white with shock and fear.

"It is a madness, yes," answered García. "It is the madness of the heart, a terrible death."

"What are we to do?" Memo asked in a stricken voice. "I do not know what to do. What of my boys on the mountain? Where are the others? I do not know what will happen to us."

García looked at Beto, who was standing as still and empty as a deserted house, while the girl lay cowering and sobbing on the floor. *I can do nothing,* García thought. *And if I try, it will only enrage him more.* He turned to Memo.

"Come, I will help you," he said, and the two ran from the church.

Alone, Beto looked down at the huddled flesh, the ragged sheet, on the floor before him.

"I have loved you," he said, his voice like steel. And then sobs came, choking him. "Whore! I have loved you!" And stooping, he ripped away her clothes. He laid bare the beautiful perfect body he had dreamed of. And he took her there on the altar beneath the dead plaster Christ, tearing her gentle white flesh. "I have loved you," he said as he pierced her, "oh, I have loved you and you have killed me."

She bled then, at last. She bled from the loss of her virgin membrane, from the ripping of his angry penis against her soft, gentle skin, and from the clawing of his fingers at the face which had beguiled and betrayed him, while she wept and clung to him and cried, "Oh my libera-

tor, you are my liberator, you are my own love, I love you,
oh no because I love you . . ."


García was almost at the gate of the courtyard when
he remembered Rita. He went to her room, but the door
stood open and the room was empty. As he turned away,
he saw the priest, the ridiculous Father Herrera, nodding
to himself as if pleased.

"You are leaving, señor?" the stupid man asked
pleasantly, a note of amusement in his voice. García was
baffled.

"Sí, Padre, we are leaving. You should be also, if
you value your life in this world. Or are you so ready to go
to your god?"

Herrera said genially, "Oh, but it is you who are
going, señor, though not to God I expect. And I will have
back my kitchen and my criada and my church and my
village."

García began to laugh harshly, and then he could not
stop. He laughed as he had not laughed in months, years,
in his whole life. He laughed until his belly ached and his
eyes ran. He laughed as if he would never laugh again.

"Oh yes, Padre," he said at last. "You will have
back your church, with its courtyard full of weapons and
munitions, a plain sign to the army. Soon you will have a
bombed-out crater, and you yourself will be lying a heap
of cracked and burned bones at its center. I salute you,
Padre, for you have won." And he turned and walked
away.

"But—" Father Herrera called after him stupidly.

"The army will not harm me. Why should they? I am loyal!"

"I believe you, Padre," García said without turning back. "I salute you!" And he walked on, laughing. As he crossed the courtyard, everything was clear to him. Fate had taken them all into its fatal, unfathomable web and now held them there, helpless, with no power save over their own souls. All that remained was to act in beauty at the last.

By the gate, Memo waited for him anxiously. "Come quickly," he said to García. "You said you will help me."

"Go ahead of me, muchacho. Go to your boys and I will come soon."

"But you said—" Memo's voice was sharp and high with fear.

"I say now that you must go. Do you wish to be a man at this moment? Or a boy? It may be that this is the last thing you will do. How will you go—as a man or a child?"

Memo could not answer. He did not know the answer. He did not decide. He turned and walked, blindly, back again to the mountain.

García sighed, a sigh so deep it seemed to come from the bottom of the world. Very little time remained. Already his ears seemed—almost—to hear planes approaching. *I cannot leave them,* he thought, remembering Beto lost in the madness of violence, and the girl, trapped in a moment too terrible to bear. *At least Beto must know the truth.* He walked across the courtyard, now utterly empty, and the sounds from the air became louder, though whether they were in his mind or in the blue, sun-filled sky, he could not tell.

Inside the church, the shadows were deep, and the

two did not see him. But García could see them, spent and weeping, huddled together on the stone floor of the sanctuary. He saw the girl, her clothes torn, her face bleeding. He saw the man, the strong, courageous Beto, his cheeks streaked with mingled blood and tears. So alike they seemed to him in that moment, so like two halves of one being, that a shock ran through him. "Roberto, mi amigo," he whispered to himself. "And Luna, the face of the moon." Now he remembered words spoken by a fire in the middle of some night, a story that Beto had told him. A family killed, a sister with the white face of the moon when it is full.

Outside, the first bomb fell, and the earth under Pasaquina trembled. The dry fountain shattered and the trees around the plaza burst into flame, like magical bushes in the wilderness. Across the courtyard, Rita lay very still, on the bed she had made for Sister Magdalena, and she gazed at the faded mural. In her mind she saw Pasaquina as it might have been once, or as it might have been someday, the bright houses, the flowers, the laughing children, the doves fluttering around the bell tower. And as she watched, a crack started at the top of the wall and broke its way down, splitting the painted bell tower, and the doves flew away on their chunks of plaster.

García felt the shaking of the earth and heard the first of the sounds that he knew would come again and again, the pounding that would break all the walls and all the people of this poor village which meant nothing to anyone and soon would cease to be. The bombs would take the dogs and the cats, the tattered awnings and the rickety tables, the crippled grandfathers and the runny-nosed babies and the plump produce vendors. The air would be filled with smoke and dust and the cries of the dying, and then all would be quiet again and the blood would dry on the broken stones and the sun would bake the mangled bodies and Pasaquina would be no more.

Before him, the two lovers were oblivious to the rain of death which was beginning. Wrapped now in the mysteries of their belonging together, they knew nothing but the cruel peace they had found at last. Their voices came to him softly. "Querida mía," one said, and "mi corazón," the other, over and over, and their words twined together so that he could not tell which was which.

*The last of their family,* García thought. *The war has finished what it began.* Then in García's heart, a song of sorrows formed itself, and he was filled with the sound, which beat as his heart beat. It sang sadly inside him and for a moment mingled with the two softly twining voices. And then García turned and was gone.

)